Portland Community

WITH

KU-168-734

£1.80

DH2

Catastrophes and lesser calamities
The causes of mass extinctions

Catastrophes and lesser calamities

The causes of mass extinctions

Tony Hallam

OXFORD
UNIVERSITY PRESS

OXFORD
UNIVERSITY PRESS

Great Clarendon Street, Oxford OX2 6DP

Oxford University Press is a department of the University of Oxford.
It furthers the University's objective of excellence in research, scholarship,
and education by publishing worldwide in

Oxford New York

Auckland Bangkok Buenos Aires Cape Town Chennai Dar es Salaam
Delhi Hong Kong Istanbul Karachi Kolkata Kuala Lumpur Madrid Melbourne
Mexico City Mumbai Nairobi São Paulo Shanghai Taipei
Tokyo Toronto

Oxford is a registered trade mark of Oxford University Press
in the UK and in certain other countries

British Library Cataloguing in Publication Data
Data available

Library of Congress Cataloging in Publication Data
Data available

ISBN 0-19-852497-8

1 3 5 7 9 10 8 6 4 2

Typeset by Footnote Graphics Limited, Warminster, Wiltshire
Printed in Great Britain by
Biddles Ltd, King's Lynn, Norfolk

Preface

When I accepted an invitation from the Oxford University Press to write a popular account of mass extinctions, I had two things primarily in mind. Firstly, I had become somewhat exasperated over the years, as had many of my colleagues, by the unbalanced and over-sensationalized treatment of the subject of mass extinctions by the media, including respectable broadsheet newspapers and television channels. This was of course the direct consequence of the remarkable discoveries of an iridium anomaly and shocked quartz at the Cretaceous–Tertiary boundary, apparently coinciding with the extinction of the dinosaurs. This led to the interpretation of a spectacular deterioration of the global environment as a consequence of the impact of an asteroid about 10 kilometres in diameter. The combination of dinosaur extinction and asteroid impact has proved irresistible to science journalists, who are always under pressure from their editors to produce sensational material that will interest the general public.

As a direct consequence of this research on Cretaceous–Tertiary boundary strata, there has been a tendency among some very able scientists, both within and outside the Earth-science community, to ascribe many or even all catastrophic mass-extinction events to the impact of asteroids or comets, and much attention has been given to the prospect of future

Armageddon induced by phenomena from outer space. All this has meant that events produced by changes solely confined to our own planet have been underplayed by both the public and many otherwise well-informed scientists. This book is intended to redress the balance and put impacts within the context of a number of purely Earth-bound events that have evidently affected the biosphere severely on a number of occasions in the geological past.

The second matter concerns the ignorance of the general public, however intelligent or well educated they may be in other respects, about how geologists and palaeontologists study mass-extinction events from data they extract from the geological record across the world. I have thus tried to give at least an elementary idea about how they go about this, and how ideas or speculation are put to the test, which is of course the essence of the scientific method. I thus hope to improve understanding to some degree, while remaining aware that in any subject there are degrees of understanding. For those who wish to learn more, there is a list of references at the end of the book with suggestions for further reading. I hope also that, besides interested members of the general public, professionals and students in the Earth sciences, together with some biologists, who wish to know more about mass-extinction studies, may derive some benefit from the book.

There are many people I could thank, but I shall confine myself to mentioning just a few. I derived great stimulation from various conferences across the world from the mid-1980s to the early 1990s, when the debate on the end-Cretaceous extinctions was at its height, and I am rather proud of the fact that I have maintained good personal relations with some of the leading antagonists, such as Walter Alvarez and Frank

Asaro on the one hand, and Charles Officer and the late Charles Drake on the other. Subsequent meetings have paid more attention to other mass-extinction horizons, most notably that at the Permian–Triassic boundary. Although I have sometimes disagreed strongly with Dave Raup, I have always appreciated and have, I hope, sometimes benefited from the lucidity of his thinking on many topics. Mike Benton has exhibited an impressive and rather rare mixture of focusing on critical detail and having an awareness of the big picture. I have greatly appreciated the work of Richard Fortey and the late Steve Gould, not just for the general excellence of their popular science writing but for their free use of personal anecdote to enliven the text, which has influenced my own style. Most of all I thank my good friend and colleague Paul Wignall for his stimulating company and wise thoughts. Bruce Wilcock proved a most percipient copy editor, for which I am duly grateful. Finally, I am greatly indebted to June Andrews for her invaluable secretarial help.

A.H.
Birmingham
February 2003

Contents

Note

In this book 'billion' denotes a thousand million (1000,000,000 or 10^9); 'trillion' a million million (1,000,000,000,000 or 10^{12}).

List of illustrations

In search of possible causes of mass extinctions

When the subject of extinctions in the geological past comes up, nearly everyone's thoughts turn to dinosaurs. It may well be true that these long-extinct beasts mean more to most children than the vast majority of living creatures. One could even go so far as to paraphrase Voltaire and maintain that if dinosaurs had never existed it would have been necessary to invent them, if only as a metaphor for obsolescence. To refer to a particular machine as a dinosaur would certainly do nothing for its market value. The irony is that the metaphor is now itself obsolete. The modern scientific view of dinosaurs differs immensely from the old one of lumbering, inefficient creatures tottering to their final decline. Their success as dominant land vertebrates through 165 million years of the Earth's history is, indeed, now mainly regarded with wonder and even admiration. If, as is generally thought, the dinosaurs were killed off by an asteroid at the end of the Cretaceous, that is something for which no organism could possibly have been prepared by normal Darwinian natural selection. The final demise of the dinosaurs would then have been the result, not of bad genes, but of bad luck, to use the laconic words of Dave Raup.

In contemplating the history of the dinosaurs it is necessary to rectify one widespread misconception. Outside scientific circles the view is widely held that the dinosaurs lived for a huge slice of geological time little disturbed by their environment until the final apocalypse. This is a serious misconception. The dinosaurs suffered quite a high evolutionary turnover rate, and this implies a high rate of extinction throughout their history. Jurassic dinosaurs, dominated by giant sauropods, stegosaurs, and the top carnivore *Allosaurus*, are quite different from those of the Cretaceous period, which are characterized by diverse hadrosaurs, ceratopsians, and *Tyrannosaurus* (**Figs. 1.1a + b**). Michael Crichton's science-fiction novel *Jurassic Park*, made famous by the Steven Spielberg movies, features dinosaurs that are mainly from the Cretaceous, probably because velociraptors and *Tyrannosaurus* could provide more drama. The implication of all this is that while there is no doubt that the dinosaurs suffered a major catastrophe at the end of their reign on Earth, it need not necessarily have been significantly more severe than a number of other events throughout their history. Unfortunately the fossil record for the dinosaurs is so patchy and limited that it is difficult at present to say much of note about such events.

Interest in the subject is reflected in the numerous hypotheses that have been put forward over the years to explain the extinction of the dinosaurs at the end of the Cretaceous period. Alan Charig, a former curator of fossil reptiles in the Natural History Museum, London, reckoned that he had discovered more than ninety, most of them more or less fanciful. They include climatic deterioration, disease, nutritional problems,

Fig. 1.1a Two dinosaurs of the Jurassic period.

parasites, internecine fighting, imbalance of hormonal and endocrine systems, slipped vertebral discs, racial senility, mammals preying on dinosaur eggs, temperature-induced changes in the sex ratios of embryos, the small size of dinosaur brains (and consequent stupidity), and suicidal psychoses. Perhaps the most fanciful of all appeared in the late 1980s in a letter to the *Daily Telegraph*, from a scientist respectable enough in his own field. He thought that the dinosaurs had died out as a consequence of an AIDS infection induced by viruses introduced from outer space. One of my favourites relates the demise of the dinosaurs to the decline in the Cretaceous of the naked seed plants, or gymnosperms, at the expense of the flowering plants, the angiosperms. Surviving gymnosperms, which include cycads (palm-like trees) and conifers, commonly

Fig. 1.1b Dinosaurs of the Cretaceous period.

contain fluids with renowned purgative properties. The implication, then, is that the herbivorous dinosaurs died of constipation. The problem with this hypothesis is that the main floral turnover took place about 35 million years too early. I used to recount this hypothesis in a course on the elements of historical geology that I gave to a class of first-year geography students at Oxford. It wasn't just for the usual lecturer's trick of raising a laugh (which it always did) to keep the attention of the class, but to make a serious scientific point, which was that, unlike most of the *ad hoc* speculations that have passed as extinction hypotheses, this one could be tested against the stratigraphic record – and found wanting. Two other consequences are worth recounting. About twenty years after he had attended my course, a man approached me at a conference. He was polite enough to tell me how much he had enjoyed the lectures, but he then rather spoiled the effect by admitting that the dinosaur joke was the only thing about them that he had remembered. The other outcome is that, as a result of a short article I wrote repeating the story, two popular science books have been published in the United States that attribute the authorship of the 'constipation hypothesis' to me . . .

As the spread of an epidemic disease has so often been invoked for the extinction of the dinosaurs (and indeed of other groups), the subject of disease as an extinction mechanism needs to be addressed. The onset of the Black Death throughout Europe in the mid-fourteenth century undoubtedly ranks as a catastrophe by anyone's reckoning, but the human race did not become extinct as a consequence. The same applies to pandemics, such as the outbreak of influenza that swept through Europe and America in 1918, reaching the remote wastes of Alaska and the most isolated of island communities. It is estimated that half the world's population was infected, and that of those infected one in twenty died. Most of the fatalities were among teenagers and young adults. This compares with a death rate of one per thousand in other flu pandemics, in which most of those who die are either very young or very old. The 1918 'Asian' flu virus was evidently one of unusual virulence but, catastrophe though it undoubtedly was, our species survived.

A third example is the smallpox virus that was carried to the Americas, the whole of Africa, Australia, and New Zealand by European settlers and invaders. In each of these territories the indigenous population suffered terribly because of their total lack of resistance. In some places the disease killed up to half the population and had a significant influence on world history. Thus, probably a third of the Aztecs were killed by smallpox, allowing Cortez an easy victory. The Aztec peoples did not, however, become extinct and today their descendants dominate the population of Mexico.

The key point, of course, is that even in the most catastrophic epidemics or pandemics caused by viruses or bacteria a proportion of the infected population either already possesses or in the course of time acquires immunity through adaptation. Thus, whereas the Europeans had lived with the smallpox virus for centuries, it was new to peoples in other parts of the world. Within Europe there have been repeated outbreaks of the plague throughout the course of history, but after the Black Death its significance as a mass killer was in decline long before there were widespread improvements in housing or sanitation.

A striking illustration of adaptation comes from the man-made epidemic of myxomatosis in rabbits. The myxomatosis virus, which naturally infects Brazilian rabbits and causes them little harm, was introduced into Australian rabbits (which were originally introduced from Europe) in 1950 in a deliberate attempt to control their ever-increasing numbers. The virus initially had the devastating effect that was intended; in the first year it killed 99.8 per cent of infected rabbits, but the effect soon wore off, and seven years later only 25 per cent of infected rabbits were dying. Now, half a century later, rabbits are back to full strength in Australia. Since the generation time of a rabbit is only 6–10 months, compared to 20–30 years for humans, in a similar situation it would take us 120–150 years to adapt to a new killer virus.

It follows that it is extremely unlikely, indeed virtually impossible, for a particular disease, however devastating, to cause a species to become extinct. The situation is even more pronounced for mass extinctions, because it is unusual for diseases to cross species barriers freely. Despite the current concern about the possibility of BSE being transferred from cattle to humans in the form of new variant CJD, and the likelihood that the AIDS virus was originally transmitted to humans in Africa from apes or monkeys, the groups in question are all mammals, and hence are closely related in an evolutionary sense. Microbial transfers of this kind are generally perceived as exceptional. Disease can thus be ruled out as a plausible mechanism to account for mass extinctions, which by definition have simultaneously affected a whole variety of organisms, both terrestrial and marine, of widely differing biology. The clear implication is that such extinctions must have involved deleterious changes in the physical environment. There is indeed no serious argument on this point among the scientists who study mass extinctions, but plenty of argument about just what these changes were, and about the most likely ultimate cause.

The next point that needs to be established is that, to be really effective, the environmental changes must have been on a global scale. If they were merely regional, the areas of refuge for the survivors would have been far too extensive, and they might subsequently have expanded their geographical range once conditions had ameliorated. Furthermore, the environmental deterioration could have been effective in two ways. It could have been short-lived but so severe that only a limited proportion of organisms survived. Alternatively it might have been less severe but more sustained in time so that, to quote T. S. Eliot's lines in *The Hollow Men*, organisms died out 'not with a bang but a whimper'. Clearly this is highly relevant to the subject of how catastrophic the mass-extinction events were, and the possible causal mechanism.

There are in fact only a few geological phenomena that can plausibly be invoked. The problem is to discriminate between them: one particular phenomenon can have multiple environmental effects on a global scale. Impacts by comets or asteroids, sea-level changes, and volcanism can all affect climate, and climatic change can affect sea level and the degree of oxygenation of the ocean water. Disentangling the critical factor can be difficult, and some people have been tempted to throw up their hands and admit a role for every factor that might be relevant. This can be called the *Murder on the Orient Express* scenario, after the celebrated Agatha Christie whodunnit in which the final resolution turns out to be that everybody did it. Other writers have promoted one factor as the dominant, if not the only, cause of all mass extinctions, both major and minor. The approach preferred here is to attempt to steer a course, Odysseus-like, between the Scylla of one overriding cause and the Charybdis of an unresolved multiplicity of causes.

A historical science like geology suffers from the limitation that it is not amenable to experimental test. Indeed, Karl Popper, the philosopher who is well known among scientists for his criterion of falsifiability, argued that for this reason the historical sciences (which include evolutionary science) hardly rank as proper science at all. Popper's falsifiability criterion has, however, been attacked by other philosophers as being over-simplistic, and a consensus has emerged that the true distinction between science and myth is that science is amenable to objective testing of its hypotheses, whether by experiment or by observation. The types of observation that are relevant to the study of mass extinctions are mainly concerned with the association with other phenomena that can be recognized in the stratigraphic record. These topics provide the subject matter for later chapters. Before that, however, we need to consider the historical background to thought on extinctions and catastrophes. We shall then be better able to evaluate the significance of modern research on these events.

2

Historical background

Georges Cuvier has not been treated with much respect in the English-speaking world for his contributions to the study of Earth history. Charles Lyell is thought to have effectively demolished his claims of episodes of catastrophic change in the past, and it is only in the past few decades, with the rise of so-called 'neocatastrophism', that a renewed interest has emerged in his writings, which date from early in the nineteenth century.

Cuvier was a man of considerable ability, who quickly rose to a dominant position in French science in the post-Napoleonic years (**Fig. 2.1**). Though primarily a comparative anatomist, his pioneer research into fossil mammals led him into geology. He argued strongly for the extinction of fossil species, most notably mammoths, mastodons, and giant sloths, at a time when the very thought of extinctions was rather shocking to conventional Christian thought, and linked such extinctions with catastrophic changes in the environment. This view is expressed in what he called the 'Preliminary Discourse' to his great four-volume treatise entitled *Recherches sur les Ossements Fossiles* (Researches on fossil bones), published in 1812. This

Fig. 2.1
Georges Cuvier (1769–1832).

extended essay was immensely influential in intellectual circles of the western world, was reissued as a short book, and was repeatedly reprinted and translated into the main languages of the day. It became well known in the English-speaking world through the translation by the Edinburgh geologist Robert Jameson (1813), who so bored the young Charles Darwin with his lectures that he temporarily turned him off the subject of geology. According to Martin Rudwick, who has undertaken a new translation which is used here, Jameson's translation is often misleading and in places down-right bad. It was Jameson's comments rather than Cuvier's text that led to the widespread belief that Cuvier favoured a literalistic interpretation of Genesis and wished to bolster the historicity of the biblical story of the Flood.

The English surveyor William Smith is rightly credited with his pioneering recognition of the value of fossils for correlating strata, which proved of immense importance when he produced one of the earliest reliable geological maps, of England and Wales, but the more learned and intellectually ambitious Cuvier was the first to appreciate fully the significance of fossils for unravelling Earth history. Whatever his attractive qualities, modesty was not one of them, and he hoped to do for the dimension of time what Newton and his French compatriot Laplace had achieved for space. In essence he laid down a research programme based on the use of fossils found in successions of rock strata, which he treated as historical documents, at a time when geologists were still trying to establish correlations by means of the rocks themselves. His programme entailed studying the youngest and most familiar rocks and fossils first, and working backwards through time. His own field experience was effectively limited to the relatively young Quaternary and Tertiary strata of the Paris Basin in northern France, where his research was done in collaboration with his Parisian colleague Alexandre Brongniart. It was Brongniart who was able in 1821 to demonstrate the presence of Cretaceous fossils at an altitude of about 2000 metres in the Savoy Alps and of fossils resembling those of the Tertiary deposits of the Paris Basin high in the Vicentine Alps. This work clearly indicated for the first time that some mountains must, in geological terms, be young in age.

In his Preliminary Discourse, Cuvier pointed out that the lowest-lying strata we see are full of marine shells, indicating that the sea had invaded the plains and stayed there peacefully for a long time. At the feet of mountains, however, the strata become tilted, and the species they contain are different from those found in younger rocks. These tilted beds form the crests of what in Cuvier's time were called the 'secondary mountains', and plunge below the horizontal beds of hills that form their feet. Some unspecified cause had broken, tilted, or otherwise disturbed them. The catastrophe that made these beds oblique had also thrust them above sea level. Fossil species and even genera changed with the successive beds or strata. In the middle of the marine beds there are other beds containing only terrestrial and freshwater plants or animals. Thus the successive catastrophes of our planet have caused alterations of marine and terrestrial conditions. The catastrophes that led to such changes have been sudden. Cuvier chose the fossil mammoths, which had then recently been discovered, as an example. Because many such mammoths known from northern Siberia are well preserved today, there must have been a very rapid cooling of the climate to preserve them. The tearing and upheaval of beds that happened in earlier catastrophes were as sudden and violent as the latest one. Masses of debris and rolled stones (evidently what we now call conglomerates) are found between the 'solid' beds, and attest to the force of the movements that these upheavals generated in the body of water.

Thus life on earth has often been disturbed by terrible events: calamities which initially perhaps shook the entire crust of the earth to a great depth, but which have since become steadily less deep and less general. Living organisms without number have been the victims of these catastrophes. Some were destroyed by deluges, others were left dry when the sea bed was suddenly raised; their races are even finished for ever, and all they leave in the world is some debris that is hardly recognisable to the naturalist.

In his commentary on Cuvier's writings Rudwick is at pains to point out that his later reputation as a highly speculative 'theorist of the Earth', of the type quite common in the eighteenth century, is an unfair one for he had repeatedly criticized that whole genre as involving a morass of ill-founded conjectures. Both his fossil anatomy and geology were based on careful 'actualistic' comparisons with living animals and present-day geological processes. He invoked catastrophes only when, in his opinion, such processes were clearly inadequate to explain what could be observed. In his telling phrase, 'the thread of operations is broken'. On the other hand, although Cuvier's stratigraphic research was at first empirically based, his concluding inference of global revolutions of a catastrophic nature, sufficient to wipe out whole faunas and floras, was a gross extrapolation from a very limited field area, which rendered him vulnerable to the question of just how widespread, contemporary, and catastrophic the changes were.

Popular mythology, at least among generations of geology students, has it that Cuvier and his fellow 'catastrophists' were put to rout by a certain knight in shining armour called Charles Lyell (**Fig. 2.2**). As usual, the true story is more complicated; and it is perhaps not so flattering to Lyell. The work with which Lyell made his name was a huge treatise published in three volumes between 1830 and 1833. It is generally known as the *Principles of Geology*, but its full title presents the essence of what he was trying to do: *Principles of Geology, Being an Attempt to Explain the Former Changes of the Earth's Surface by Reference to Causes Now in Operation*. Lyell has often been credited for being able to infer the immensity of geological time from a study of natural phenomena operating at the present day (a decidedly heretical view for those many contemporaries of his who were orthodox

Fig. 2.2
Sir Charles Lyell (1797–1875).

Christians). In fact the person who originally proposed this idea was another Scotsman, James Hutton, who has much the greater claim to being one of the true fathers of geology.

As early as 1788 Hutton wrote the following key sentences. Despite the antique flavour of the language the meaning comes through quite clearly.

> In examining things present, we have data from which to reason with regard to what has been; and, from what has actually been, we have data for concluding with regard to that which is to happen hereafter. Therefore, *upon the supposition that the operations of nature are equable and steady*, we find, in natural appearances, a means of concluding a certain portion of time to have necessarily elapsed, in the production of those events of which we see the effects.

Note the prime assumption of the phrase here italicized, from which Hutton went on to introduce the notion of indefinite time. It is the essence of Lyell's actualistic method, but Lyell never gave adequate acknowledgement to his great Scottish predecessor. In 1823 Lyell paid a lengthy trip to Paris where he met, among others, Cuvier, Brongniart, and Humboldt. The French geologist who influenced him most was, however, Constant Prévost, and Hutton travelled extensively with him on several occasions. Prévost's researches in the Paris Basin had demonstrated how a succession of small changes could convert a sea into a freshwater lake. He argued persuasively that the Paris Basin in Tertiary times was an inlet of the sea, with the salinity diminishing eastwards as the distance from the open marine connection increased. This work raised serious doubts about the drastic character of environmental change claimed by Cuvier, but Prévost's attempts to challenge the great man's authority in France met with little success.

Lyell's belief in the heuristic value of the actualistic method stems directly from Hutton, but in promoting a steady-state, uniformitarian 'system' Lyell goes further than Hutton:

> There can be no doubt, that periods of disturbance and repose have followed each other in succession in every region of the globe, but it may be equally true, that the energy of the sub-terranean movements have been always uniform as regards the whole earth. The force of earthquakes may for a cycle of years have been invariably confined, as it is now, to large but determinate spaces, and may then have gradually shifted its position, so that another region, which had for ages been at rest, became in its turn the grand theatre of action.

Now Hutton had written '. . . we are not to limit nature with the uniformity of an equable progression . . .' – hardly a uniformitarian sentiment. He also envisaged catastrophic uplift, whereas Lyell saw no need to invoke such drastic effects. In his view, erosion is balanced by deposition, subsidence by elevation, in a system that was in a steady state with fluctuations about a mean. He shared with Hutton a belief in a relationship between earthquake activity, volcanicity, and uplift, and in a rejection of any notion of progression or directionality through time. Lyell's steady-state model of Earth evolution was at variance with the directionalism favoured by the leading English geologists of the time. Because of a general belief in a gradually cooling Earth, they doubted that 'the energy of the subterranean movements has been always uniform as regards the whole earth.' By invoking the operation of greater forces in the past, these people have been somewhat unfairly dubbed catastrophists. Consequently Lyell had few supporters initially, and one of these was Charles Darwin (**Fig. 2.3**). As a young man Darwin showed great promise as a geologist, but he subsequently became sidetracked. In chapter 10 of *On the Origin of Species* he makes the following categoric statement: 'He who can read Sir Charles Lyell's grand work on the Principles of Geology, which the future historian will recognise as having produced a revolution in natural science, and yet does not admit how vast have been the past periods of time, may at once close this volume.' Lyell's gradualism indeed became the key to Darwin's biological gradualism, and catastrophic changes were not called for in either the geological or biological record.

When we turn to Lyell's comments on catastrophism in *The Principles of Geology* we uncover a curious fact. The catastrophism that Lyell dismisses is that of the ancient Hindus or ancient Egyptians. 'Universal catastrophes of the world, and

Fig. 2.3
Charles Darwin (1809–1882). This portrait was drawn when Darwin was better known as a geologist than as a biologist.

extermination of organic beings, in the sense that they were understood by the Brahmin, are untenable doctrines.' Nowhere is there as much as a mention of Cuvier's catastrophism, although there are many respectful references to Cuvier on other subjects elsewhere in the treatise. There can be no question that such a well-read scholar as Lyell was familiar with Cuvier's Preliminary Discourse, so why the omission? Could it be that Cuvier's prestige was so great that it would have been unwise to attack him directly, and 'the Brahmin' was a coded reference?

At any rate, in the middle of the century Darwin felt free in *The Origin* to state with assurance that

> The old notion of all the inhabitants of the earth having been swept away at successive periods by catastrophes is very

generally given up, even by those geologists . . . whose general views would naturally lead them to this conclusion. On the contrary, we have every reason to believe, from the study of tertiary formations, that species and groups of species gradually disappear, one after the other, first from one spot, then from another, and finally from the world.

Darwin was well aware at that time that there were what he called 'imperfections' in the geological record, so that it was difficult to recognize successions of fossil species that could plausibly be held to represent evolutionary series. He explained this deficiency in two ways. Firstly, because fossils would be preserved only in zones of subsidence rather than of uplift or stasis, there were numerous gaps in the stratigraphic record in given areas. Secondly, migration of species could have taken place from other areas. One could add that the types of organisms preserved would vary according to the environment. By implication, although once more his name was not mentioned, analyses such as had been performed by Cuvier in the Tertiary deposits of the Paris Basin were quite inconclusive so far as invoking catastrophes was concerned.

Lyell remained on friendly terms with Darwin, whom he regarded in some respects as his protégé, but found it difficult to come to terms with his theory of evolution by natural selection, with its clear implication of organic change, and apparent progression, through time. Lyell was particularly troubled by the implications of the theory for our own species, whose history belonged, he thought, to 'the moral sphere'. His conversion therefore came about with reluctance and late in life. His uniformitarianism also came under strong attack from Kelvin (**Fig. 2.4**), who made an estimate of the Earth's age based upon cooling from a molten mass. Kelvin's figure,

Fig. 2.4
William Thomson, Lord Kelvin
(1824–1907).

originally between 20 million and 40 million years, was eventually reduced to a mere 24 million years by the end of the nineteenth century. With the discovery of radioactivity in rocks shortly afterwards, and the emergence of radiometric dating at the start of the twentieth century, this figure was quickly abandoned together with Kelvin's theoretical assumptions. The Earth's age was then extended to hundreds of millions of years and then to a few thousand million years. Lyell's call for almost indefinite time seemed subsequently a reasonable approximation, because evidently geologists had vast amounts of time to play with, on strong scientific grounds. Any cooling of our planet through time could therefore effectively be disregarded for the time for which there was a good stratigraphic record, and Lyell's steady-state model of

Earth evolution seemed like a reasonable approximation. Generations of geology students were therefore indoctrinated with Lyellian gradualism and were taught to repeat as a catechism the rather question-begging and banal phrase, 'The present is the key to the past.'

Modern developments

In the latter part of the twentieth century a new school of thought emerged in geology that has been dubbed 'neo-catastrophism', as opposed to the gradualism of Lyell and Darwin. Its adherents embrace a punctuated view of geo-logical and biological history. Thus the stratigraphic record has been compared by Derek Ager to the traditional life of a soldier: long periods of boredom interrupted by moments of terror. Those who study sediments and sedimentary rocks have become increasingly interested in the rare intense storm or turbidity current, which may have more significant erosional and depositional consequences every few decades than the modest everyday activities more amenable to direct observation. Lyellian extrapolation of such processes to larger events, simply adding the time dimension to increase magnitude by cumulative effect, may often not be justifiable. There has been a parallel shifting of thought among geomorphologists. Thus the once-heretical conclusion that the barren channelled scab-lands of eastern Washington State in the United States were eroded by catastrophic floods has in recent years obtained widespread acceptance. There is even a fairly new 'catas-trophe theory' in mathematics, put forward by the Frenchman

René Thom, which had quite a vogue in the 1970s before being rather eclipsed by chaos theory. It postulates that a gradually changing *cause* may be reflected, not in a gradually responding *effect*, but by a sudden shift from one stable state to another. This could well be relevant to crustal tectonics, in which the slow accumulation of stress could be relieved episodically by violent and destructive earthquakes.

The question is just as relevant to organic evolution. Darwin was so wedded to the notion of gradualistic change, which he derived from his hero Lyell, that he was positively embarrassed by the numerous gaps and lack of transitional forms in the fossil record. His explanation, that this is a consequence of the extreme imperfection of the stratigraphic record, in which many intervals of time are unrepresented, has become progressively less plausible as the years have passed. Everexpanding research activity across the world, including the ocean floor as well as the continents, has filled out the stratigraphic record to an enormous extent, but many of the fossil discontinuities have persisted. Niles Eldredge and Steve Gould grasped the nettle by proposing that the conventional gradualistic view of species change through time was in error. They put forward their well-known alternative, termed 'punctuated equilibria', in which long periods of morphological stasis are interrupted by brief, geologically 'instantaneous' episodes of speciation, when significant genetic and consequently morphological change takes place. Whole groups of species may exhibit such stasis before they change together, a phenomenon that has been termed 'coordinated stasis'. The speciation event is presumably related to significant environmental change, after a long period of boredom, to use Ager's phrase. This leads us on to the subject of mass extinction, on both large and small scales. Ironically, while Ager remained until his death an enthusiastic neocatastrophist as regards geological processes, he persisted in his belief that mass extinctions, which perhaps could have provided some of the best evidence in support of his position, were gradualistic phenomena. It is time therefore to enquire more closely into what we mean by catastrophic mass extinctions and how we can recognize them in the stratigraphic record.

3

Evidence for catastrophic organic changes in the geological record

If asked what they understood by the word 'catastrophe', most people would probably agree that it was something big. bad, and sudden, and involved damage to organisms. In the natural world today, perhaps the most striking catastrophes result from major earthquakes, in which thousands of people can be killed within minutes. Going back through human history, we allow for greater stretches of time. Thus, in the middle of the fourteenth century, over a period of five years, an estimated one-third of the European population died directly as a result of catching the plague: the 'Black Death'. By any reckoning this ranks as a catastrophe. It had a dramatic effect on European society for many years. When we extend our consideration to geological time, in which it is routine to deal with changes taking place over millions of years, events lasting only a few thousand years may be regarded as catastrophic if the contrast with the 'background' is sharp enough.

Various definitions have been proposed for a mass extinction. A conveniently concise if imprecise one that I favour is that it is the extinction of a significant proportion of the world's living animal and plant life (the *biota*) in a geologically insignificant period of time. The imprecision about the extent of an extinction can be dealt with fairly satisfactorily in particular instances by giving percentages of fossil families, genera, or species, but the imprecision about time is more difficult to deal with. An important question about mass extinctions is to assess how catastrophic they were, so we also require a definition of 'catastrophe' in this context. One thought-provoking attempt at such a definition is that a catastrophe is a perturbation of the biosphere that appears to be instantaneous when viewed at the level of detail that can be resolved in the geological record.

At this point more needs to be said about the nature of the geological record. The material that geologists and palaeontologists deal with occurs in the layered successions of sedimentary rocks, mainly sandstones, shales, and limestones, that can clearly be observed in good rock exposures, either natural ones, as in coastal cliffs or mountains, or artificial ones, as in quarries or borehole cores. Although the principle of stratal succession – that the higher-lying strata were the younger – had first been enunciated in the late seventeenth century, it was not until the early nineteenth century that it was fully appreciated that here was a record of Earth history if we could interpret it correctly. The prime need was to establish a relative timescale, whereby we could correlate rocks across the world and thus establish the contemporaneity of events. The establishment of such a timescale, following the pioneer research on the stratigraphic use of fossils by William Smith in England and by Cuvier and Brongniart in France, was one of the great scientific achievements of the nineteenth century.

Figure 3.1 shows the geological eras and periods, based upon the fossil succession, that are accepted today. It may be

Era	Period	Epoch	Age in 10^6 years	Major extinction events
Cenozoic	Quaternary (Sub-period)	Holocene	0.01	
		Pleistocene	2	
		Pliocene	5	
		Miocene	25	
		Oligocene	38	
		Eocene	55	
		Palaeocene	65	*
Mesozoic	Cretaceous		144	
	Jurassic		200	*
	Triassic		250	*
Palaeozoic	Permian		286	
	Carboniferous		360	*
	Devonian		408	
	Silurian		438	*
	Ordovician		505	
	Cambrian		545	

Fig. 3.1 Timescale for the Phanerozoic eon. Ages are shown in millions of years. The asterisks signify the stratigraphic location of the 'big five' mass extinctions.

useful to indicate the origin of the names, which every geology student is obliged to learn by heart. They were proposed by British, French, and German stratigraphers, mostly early in the nineteenth century. 'Cambrian' is derived from the Roman name for Wales; 'Ordovician' and 'Silurian' from the names of ancient British tribes in the Welsh region who fought the Romans. 'Devonian' obviously comes from the English county, and 'Carboniferous' from the fact that its rocks contain the most important coal deposits of western Europe, the fuel of the Industrial Revolution. 'Permian' is taken from the town of Perm in the Ural Mountains of Russia, and 'Triassic' from the threefold division of strata of that age in Germany. 'Jurassic' comes from the Jura Mountains of north-western Switzerland and south-eastern France, and 'Cretaceous' from the Latin for chalk (*creta*), because this is the dominant rock type of this period in western Europe. Younger strata have been treated somewhat differently. One of Lyell's more lasting contributions was to subdivide these strata according to the proportion of living species, as they were known in the early nineteenth century, the proportion diminishing with age. Thus 'Eocene', 'Miocene', and 'Pliocene' are derived respectively from the Greek for dawn (*eos*), less (*meion*), and more (*pleion*). 'Palaeocene', 'Oligocene', and 'Pleistocene' were added by others later.

Initially the various periods and systems (the rock units corresponding to the time units) were grouped together as Primary, Secondary, Tertiary, and Quaternary, but in the middle of the nineteenth century John Phillips, Professor of Geology at Oxford, proposed three other terms that quickly became generally accepted to designate geological eras. 'Palaeozoic', 'Mesozoic', and 'Cenozoic' (originally 'Kainozoic') are derived from the Greek words for *ancient, intermediate,* and *recent life*. It is interesting that even as far back as the mid-nineteenth century it was readily recognized that there were major organic changes across the boundaries of the three eras. The Cenozoic embraces the Tertiary, the only old term that is still retained, and the Quaternary, which comprises the Pleistocene and Holocene epochs. (An epoch is a subdivision of a period.) The most widely used subdivisions of the Cenozoic (Eocene and so on) are, indeed, epochs rather than periods, unlike the Palaeozoic and Mesozoic. By the twentieth century the need was felt for an extra term, Phanerozoic, derived from the Greek for 'evident life', which collectively represents the three eras. Before this, the Precambrian had for a long time been thought to be barren of fossils. Fossils are now known to exist, but virtually all of them are microscopic and so were not 'evident' to earlier generations of geologists. Not until those highly distinctive arthropods, the trilobites, first appeared in the early Cambrian, together with many other multicellular groups, was it easy to read a fossil succession that was evident to the naked eye in the field. The Precambrian fossil record as we know it today is still very poor and limited as compared with that of the Phanerozoic, and study of mass extinctions is essentially confined to the Phanerozoic.

On the finer scale at which stratigraphers work, the key subdivision is the *biozone*, characterized by a distinctive fossil, which gives it its name, or an assemblage of fossils. The best 'zone fossils', as they are called, are those that have undergone a relatively rapid turnover in time as a result of high rates of evolution and extinction. This allows for a finer degree of stratigraphic precision. Good examples are the ammonites (**Fig. 3.2a**), whose often strikingly beautiful coiled shells adorn many museums. These relatives of the squid and *Nautilus* provide the standard scheme for zoning most of the Mesozoic. Planktonic foraminifera provide another example. They are microscopic single-celled organisms with calcite shells that from late Cretaceous times onwards have formed part of the plankton, the organisms that drift passively in huge numbers in the oceans, rivers, and inland waters. For the Palaeozoic, another planktonic group, the graptolites (**Fig. 3.2b**), are important in the Ordovician and Silurian. Research in the past few decades has now established that the best zone fossils for the Palaeozoic as a whole are the conodonts, a diverse group of small swimming vertebrates preserved as microscopic phosphatic structures that represent the feeding apparatus of the animals. All these groups lived only in the sea. Marine strata are, indeed, much easier to correlate across the world than those deposited as sediments on the continents, whether in lakes and lagoons or on coastal and river floodplains. Because mammals had a high evolutionary turnover rate, and mammal teeth are the parts of the skeleton most resistant to destruction, these are the fossils that are most successfully used for Tertiary non-marine strata. For the Mesozoic and Palaeozoic, pollen and spores offer the greatest promise as zone fossils, but most forms unfortunately have relatively long time ranges.

Stratigraphers draw up more detailed versions of the table shown in Fig. 3.1 in which subdivisions of geological time are marshalled into orderly schemes, and they use the term *chronostratigraphy* (time–rock stratigraphy) for this branch of stratigraphy, which is concerned with interpreting the history of the Earth by means of the chronological sequence of its sedimentary rocks. (*Biostratigraphy* is the term used for the branch of stratigraphy that uses fossil animals and plants for relative

Fig. 3.2a+b Example of fossils of high biostratigraphic value.
(a) A Cretaceous ammonite. (b) An Ordovician graptolite.

dating and correlation. *Lithostratigraphy* is concerned with the lithological features of rocks and their spatial relationships.) In chronostratigraphic tables, biozones are grouped together into *stages*, which are grouped successively into epochs (which, however, are little used in practice), periods, and eras. Thus the Jurassic, for example, comprises about sixty ammonite zones (the number varies in different regions of the world) and eleven stages. Stages provide the international lingua franca of stratigraphers and thus it is very important for professional geologists to be familiar with them as well as with the names of the geological systems. (Students, however, tend to be reluctant to learn these names unless subjected to some kind of compulsion, something which is unfashionable in modern educational circles.) Stage names are clearly recognizable by having the suffix 'ian', the rest of the name being derived from some geographical locality. Thus for the Cretaceous nearly all the twelve stage names come from France. 'Cenomanian' comes for the Latin name for Le Mans; 'Coniacian' and 'Campanian' from the brandy- and champagne-producing Cognac (a town) and Champagne (a province). In this book stage names will be used only sparingly, but it is necessary to appreciate, for example, that the youngest stage of the Cretaceous is the Maastrichtian and the oldest stage of the Tertiary the Danian. In studying the Cretaceous–Tertiary, or K–T, boundary, then, the only hope of locating a complete stratigraphic succession is to find evidence for the presence of both stages. (Non-geologists often wonder, incidentally, why the Cretaceous is often abbreviated to K, rather than C in discussions of the end-Cretaceous extinctions. It is simply to avoid confusion with the abbreviations for Cambrian (C) and Carboniferous (C).) Recognition of continuity in a given succession is a matter of the utmost importance in all kinds of geological studies, because the presence of a hiatus or *unconformity* may significantly affect interpretation.

Since the latter part of the twentieth century, technological advances have made possible a variety of chemical and physical methods that can be used for stratigraphic purposes. Techniques such as wireline logging (which utilizes the electrical resistivity of rocks) or gamma-ray spectroscopy have proved valuable to oil companies for the correlation of borehole cores, but their use does not extend beyond a limited region, such as a particular sedimentary basin. For long-range correlation across the globe, other methods are needed. In isotope stratigraphy, variations through stratal successions in the ratios of isotopes of oxygen and strontium have proved their worth. Oxygen isotope stratigraphy is, for example, widely used for correlating marine Quaternary borehole cores across the world. The method is based on the fact that the ratio of oxygen-18 (^{18}O) to oxygen-16 (^{16}O) varies according to the temperature of the sea water. The $^{16}O/^{18}O$ ratios are accurately recorded in the shells of foraminifera, whether they live at the surface or the bottom of the ocean. One of the great achievements of the past few decades has been the confirmation of the Milankovitch theory. This theory postulates that the Quaternary ice ages were controlled by cyclic variations in the Earth's orbit, which gave rise to temperature fluctuations on the Earth's surface. In this work oxygen isotope stratigraphy played a critical role. Among physical methods, magnetostratigraphy has proved a valuable method of correlation across the world, mainly for Cenozoic rocks. The method is based on the succession of polar reversals in the Earth's magnetic field, which can be detected in sedimentary rock successions by sensitive magnetometers. All these chemical and physical methods are ultimately dependent, however, on biostratigraphy for calibration, and are especially useful when the appropriate fossils are absent.

Determinations of the absolute ages and durations of events in terms of numbers of millions of years are obviously of vital importance in learning more about the geological past. Radioactive isotopes provide the main modern dating methods. By measuring the ratio between the original isotope and the isotope formed from it (the 'parent' and the 'daughter' isotope) it is possible to determine when particular minerals in a rock became a closed chemical system. Early in the twentieth century the British geologist Arthur Holmes did pioneer work in using data from uranium and lead isotopes to establish the first geological timescale. There has since been a huge increase in the sophistication of the mass spectrometers that are employed, and in the variety of radioactive isotopes used, with a corresponding increase in both precision and accuracy of the dates established, but there have been no striking changes in the Phanerozoic timescale. Thus Holmes arrived at a date of about 500 million years for the base of the Cambrian, whereas today the figure of 540 is generally accepted.

It needs to be appreciated that Holmes's achievement in establishing a geological timescale was not a straightforward matter. Only a limited number of rocks contain the appropriate minerals with radioactive elements in sufficient quantities. Those used by Holmes were all igneous, and the relationship of igneous rocks to the enveloping strata had to be determined. Thus, a given granite may be found to have intruded strata of a certain age on the relative timescale as determined by fossils, and hence must be younger than those strata. Elsewhere the same granite may be overlain by other strata containing reworked pebbles or boulders of the granite and which must therefore be younger. Holmes also used the maximum thicknesses of strata of various geological periods across the world as a measure of the relative duration of time. Within stratal successions themselves, the only ones that can be used for dating are volcanic, and volcanic deposits are absent from many successions. There are also many complications, which cannot always be eliminated with the use of more sophisticated technology, about the interpretation of the chemical results. These commonly concern alterations to the isotope ratios after the minerals were first formed. Radioisotope dating thus has built-in uncertainties, which are normally shown graphically (as 'error bars') in presenting results.

For most of the time historical geologists are not concerned with more than relative ages, for instance in the study of ancient environments. I recall my first fieldwork as a research student, on the Dorset coast west of Lyme Regis, where a Lower Lias (basal Jurassic) succession of shales and limestones is overlain with marked unconformity by what is known as the Upper Greensand of mid-Cretaceous age, forming the cliff tops in the region, such as the celebrated Golden Cap on the coast east of Charmouth. My interest was in the environment of deposition of rocks of the Blue Lias Formation. The ammonite zonation of these strata had been undertaken in great detail, but almost nothing was known about the environmental conditions under which they had been deposited, apart from the fact that they were marine. When the occasional curious holidaymaker asked what I was doing on the rocky foreshore, I initially attempted an honest answer, no doubt in a rather pompously pedagogic way, but I soon noticed that their eyes were beginning to glaze over. So eventually I gave up and told them that I was looking for fossils. This seemed to satisfy them fully and off they went, no doubt to tell their friends and relatives what an interesting encounter they had had with a real-life fossil hunter.

One ten-year-old boy did, however, give me pause for thought. He brought me a piece of dark Blue Lias limestone that he had picked up on the foreshore and asked me how old it was. Pretending to scrutinize the specimen I cast my mind back to the Holmes timescale I had learned as an undergraduate and replied rather promptly 'about 180 million years'. He seemed suitably impressed and came back soon afterwards with a piece of yellow sandstone that had clearly fallen from the Upper Greensand cliffs high above. 'Mister,' he said, 'how old is this one then?' Back to Holmes, and the answer 'about 100 million years'. 'Hey mister,' he replied, 'how can you tell the age of rocks just by looking at them?' I often wonder what happened to that boy in later life. I think he might have made a good scientist, because he showed at an early age the right combination of intense curiosity, keen observation, and a reluctance to accept glib answers from adults.

Stratigraphic correlation on its own, essential though it is, is rather an arid subject, as generations of students have found. It is only by trying to interpret the changing environments of the past that historical geology comes to life. It is useful here to introduce an important geological term, *facies*, which is Latin for 'face' or 'aspect'. In the context of interpretating ancient environments it refers to the sum total of the characteristics of a sedimentary rock, that is both its lithological and mineralogical characters as well as the fossils it contains. 'Facies' can be used in different ways in particular contexts. Thus it can be purely descriptive, as in 'limestone facies', but it is more widely used in an interpretive sense, such as 'shallow' or 'deep marine facies'. Correct environmental interpretation is vital to understanding what went on across particular extinction horizons. Inevitably, because the matter is so important, it is the subject of many disputes. Nevertheless consensus is readily achieved in important areas. For example, fossils are useful, not just for stratigraphic correlation, but for environmental interpretation. Thus all living representatives of particular marine invertebrate groups such as corals, sea urchins, and brachiopods, together with many others, are exclusively marine. Sedimentary rocks of various ages that contain a variety of organic remains of these organisms can therefore reasonably be interpreted as having been deposited in a marine environment. Furthermore, long-extinct groups such as trilobites, because they occur in association with such fossils, can also be inferred to have had a marine habitat.

It is usually possible to go beyond distinguishing whether ancient deposits were laid down in the sea or on land and say whether they were deposited in lakes or lagoons, or in coastal and river floodplains. Information from the sediments themselves can here prove valuable. Thus, deposits laid down in shallow water commonly have distinctive characteristics that allow them to be distinguished from those deposited in deep water. These characteristics include evidence of wave action (waves do not penetrate more than a few tens of metres below the sea surface) such as symmetrical oscillation ripples in sandstones or distinctive storm deposits. Likewise, the photic zone, below which light does not penetrate, will be signified by the presence of fossil organisms that depend on light, such as bottom-living calcareous algae or reef-building corals that live in symbiotic association with algae. It is not usually difficult to infer the relative depth of the sea for a given stratal succession by using evidence from both sediments and fossils, but it is unfortunately very hard, if not impossible, to determine absolute depths with any reliability. Thick successions of relatively deep-water sediments usually signify areas of subsiding basins, while equivalent thin successions of relatively shallow-water sediments may signify adjacent swells. A change from basin to swell conditions through a succession of strata usually indicates some change in the regional tectonics underlying the sedimentation patterns.

The stratigraphic location of mass-extinction events

We saw in the previous chapter that by the beginning of the twentieth century Lyellian gradualism had triumphed over any kind of catastrophism. Palaeontologists subsequently paid little attention to the phenomenon of extinction, let alone mass extinction, for most of the century. The main handicap to any proper study was the lack of an adequate database. The situation began to improve from the 1950s onwards with the publication of the *Treatise on Invertebrate Paleontology*, a series of volumes devoted to invertebrate fossil groups that was initiated by the Geological Society of America. The first person to analyse the stratigraphic distribution of marine fossils (which represent the overwhelming majority of the fossil record) on this basis was Norman Newell of the American Museum of Natural History in New York. In a series of papers published in the 1960s he was able to recognize six mass-extinction events in the Phanerozoic. By the beginning of the 1980s the database had improved further, and Jack Sepkoski of the University of Chicago undertook, with his colleague Dave Raup, a more rigorous statistical analysis of Phanerozoic marine animal families. They were able to recognize five events in which the extinction rate significantly exceeded the background value and which hence qualified as mass extinction. Their stratigraphic location is indicated in Fig. 3.1: end-Ordovician, late Devonian, end-Permian, end-Triassic, and end-Cretaceous. The late Devonian event, the only one not at a system boundary, was at the boundary of the Frasnian and Famennian stages. All five of these events were among those listed by Newell. His sixth, at the end of the Cambrian, appears to be one of a greater number of lesser events with a high proportion of survivors, which qualify as minor mass extinctions but still mark notable perturbations of the biosphere.

The events listed by Raup and Sepkoski have become generally accepted as the 'big five', and in this book we shall concentrate on them, while noting where relevant some of the lesser events. The relative magnitudes of the five events need to be considered, as must the major organisms affected. This will entail introducing a few more geological terms. We shall also look at the terrestrial vertebrate and plant record in some instances. Table 3.1 summarizes the record for the five events. It will be seen from the table that by far the biggest event in terms of loss of diversity was that at the end of the Permian. The much-publicized event at the end of the Cretaceous shows the smallest loss.

Recording mass extinctions by fall in diversity is the most obvious thing to do from the data available, but it takes no account of the numbers of organisms involved. One species could dominate a whole ecosystem in terms of numbers of individuals, but its loss would have little effect on a total diversity count. Thus a major extinction phenomenon could be missed. Unfortunately we have to live with this limitation, because there is no reliable way of determining from the fossil record the numbers of organisms living at a given time. Variations in the abundance of fossils in a succession of strata normally relate to variations in rates of sedimentation: a high sedimentation rate dilutes the concentration of fossils. Geochemical methods can be used to tackle the question of variations of productivity, using carbon isotopes, but the results can be ambiguous. This is a matter to which we shall return later.

How catastrophic were mass-extinction events?

Determining whether a given extinction event was catastrophic or merely gradual is not a straightforward matter, because of the limitations imposed by the stratigraphic record. Consider the case represented in **Fig. 3.3**, in which a gradual extinction can appear to be catastrophic if a hiatus exists (caused either by failure of sedimentary deposition or by erosion in the region studied), leading to the absence of one or more biozones. This emphasizes the vital importance of having the best possible biostratigraphic control on the stratal successions that we study.

We have, however, to take into account the converse effect. What are known as 'range charts' record the series of occurrences of fossil species against a stratigraphic column. Such charts give a feel for the relative frequency of occurrence of

Table 3.1 The five biggest mass extinction events of the Phanerozoic

(1) End-Ordovician event (26)

The graptolites underwent the worst crisis in their history, and the conodonts were also severely affected. The dominant benthic (bottom-living) trilobites and brachiopods suffered major extinctions.

(2) Late Devonian event (Frasnian–Famennian boundary) (22)

Reef ecosystems, comprising rugose and tabulate corals and stromatoporoid sclerosponges, underwent a major crisis. Among the rest, both marine invertebrates (ammonoids, brachiopods, trilobites) and vertebrates (conodonts, agnathan and armoured placoderm fish) suffered severe extinctions. The land record is much less clear, but there could have been an important extinction among plants.

(3) End-Permian event (51)

Reefs were again seriously affected, with the final disappearance of rugose and tabulate corals. Major extinctions took place among the echinoderms, with the loss of important Palaeozoic crinoid groups and blastoids, and the bryozoans. There were also significant losses among the brachiopods, marking the end of their numerical domination of benthic communities, and foraminifera, with the final disappearance of a major, large-sized group, the fusulinids. On land, there was a significant loss of vertebrates, with over half the families disappearing, including all large herbivores. Insects suffered the greatest turnover in their long history, and there were regionally important extinctions among the plants.

(4) End-Triassic event (22)

A further crisis was undergone by reef ecosystems, with a major group of calcareous sponges almost completely disappearing. The rich and diverse Triassic ammonite faunas were almost totally wiped out, and the conodonts ended their long history. Significant extinctions took place among the now-dominant benthic bivalves, together with brachiopods. An important regional extinction event took place among terrestrial plants, but claims of a major land vertebrate extinction are disputed.

(5) End-Cretaceous event (16)

In the marine record the most striking event is the mass extinction of two planktonic groups with calcite skeletons, the foraminifera and coccolithophorids. Many benthic invertebrates were also strongly affected, and certain groups such as the ammonite and belemnite cephalopods and inoceramid and rudist bivalves finally disappeared from the record. On land the most obvious victims were the dinosaurs; other groups were much less strongly affected.

Figures in parentheses are the percentages of marine families that became extinct.

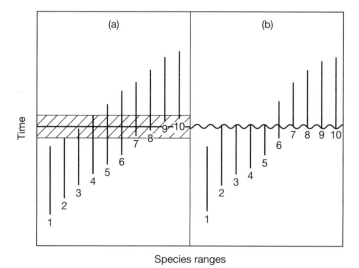

Species ranges

Fig. 3.3 Diagram to illustrate how a stratigraphic hiatus can create a false impression of a mass extinction from a continuous set of species ranges, represented by vertical bars. The left-hand portion of the diagram (a) exhibits continuous sedimentation, but in the right-hand portion (b) the band marked by oblique lines has been removed by erosion or non-sedimentation.

species; more common species will have shorter gaps between their occurrences than rarer ones. An important point to be remembered when interpreting extinctions in range charts is that the highest (i.e. latest) occurrence of a fossil species is unlikely to represent the very last individual of that species. Thus all species ranges are artificially truncated to some extent. This will tend to make mass extinctions appear more gradual than they actually were (**Fig. 3.4**). This effect is known as 'backsmearing' or the 'Signor–Lipps effect', after the two American palaeontologists who first pointed it out.

The Signor–Lipps effect is least important for the commonest species, which tend to have short gap lengths. The difference in gap length between common and rare species is the fundamental attribute that is used in several attempts to assign error bars to the last (and first) occurrences of a species. In recent years increasingly sophisticated statistical techniques have been applied to the problem, most notably with data from the Cretaceous–Tertiary (K–T) boundary mass extinction, but the subject matter presents some natural limitations. Thus it is easier to collect reliable data for tiny foraminifera, which occur in prolific quantities in small volumes of rock, than for the frequently huge but much rarer dinosaurs.

A further complication has been pointed out and analysed by another American palaeontologist, Steve Holland. He showed that a gradual mass extinction can be made to appear abrupt by rapid changes in sea-water depth, as determined by the analysis of sedimentary facies. One of the most signficant of such changes occurs with the water deepening that occurs during marine transgressions, which are characterized by marked reductions in rates of sedimentation, leading to what is called 'condensation' by stratigraphers. This will clearly

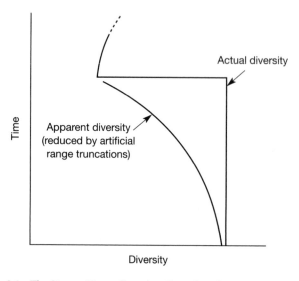

Fig. 3.4 The Signor–Lipps effect: the effect of random range truncations on an abrupt mass extinction. Artificial range truncations produce an apparently gradual extinction.

exacerbate the abruptness of organic change through time as it appears in the stratigraphic record. Thus, for those mass extinctions that are associated with rapid transgression of the sea over the land the fossil record must be searched in the expanded, near-shore stratal successions because the condensed, offshore successions may give an abrupt pattern of last occurrences that is merely the result of low sedimentation rates. Some fossil groups disappear from the stratigraphic record only to reappear much later. Such groups have been called *Lazarus taxa* by the University of Chicago palaeontologist David Jablonski, after the biblical character who returned from the dead. (*Taxon* (plural *taxa*) is a useful term for a group of organisms of any rank, such as a species, a genus, or a family.) The phenomenon is commonly due to the fact that the appropriate facies, which represents the environment favoured by the organisms, has temporarily disappeared. If this is widespread, representing a deteriorating environment across an extensive part of the world, the organisms will have retreated to a refuge of limited geographic extent, which is relatively unlikely to be sampled in the stratigraphic record.

A potential problem of quite another sort arises from the type of classification used for the fossils. Traditional taxonomy (the formal classification of organisms) is based primarily on overall degrees of morphological resemblance, leading to the distinction of *grades*. Thus, reptiles are a category clearly distinct from mammals and birds. In recent years, the more rigorous and coherent methods of *cladistic taxonomy* (from *clados*, the Greek for *branch*) have swept the board. Under this method, reptiles are *paraphyletic*; that is, they exclude their evolutionary descendants, the mammals and birds. But in cladistics degrees of affinity are based on the inferred recency of common

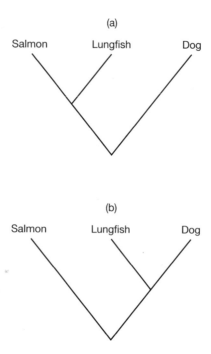

Fig. 3.5   The difference between traditional and cladistic methods of classification, illustrated by reference to three familiar vertebrates. In the traditional method (a) closeness of evolutionary affinity is determined by degree of morphological resemblance, whereas in the cladistic method (b) it is determined by recency of common ancestry.

ancestry, and paraphyletic groups are not permitted; only *monophyletic* groups or *clades*, embracing all the descendants, are allowed (**Fig. 3.5**).

In the context of extinction one can foresee a potential problem here. It is now generally accepted that dinosaurs gave rise to birds, which would make the dinosaurs paraphyletic and thus an artificial grouping. Cladists get round this by calling the dinosaurs non-avian dinosaurs, which some might consider, a little unfairly perhaps, as intellectual sleight of hand. There is a school of thought that considers that if particular fossil groups, whether monophyletic or paraphyletic, represent distinct biological entities their final disappearance from the stratigraphic record then represents significant information. While there is some truth in this, we have been taught by the cladists to be aware of the problems of *pseudoextinction*. A particular taxon may be reported as becoming extinct at a stratigraphic horizon, whereas all it has done is evolve into its descendant. The overall consequence for the study of mass extinctions is that traditional taxonomic methods have probably overestimated the extent of extinction. This means that whereas the 'big five' mass extinctions still stand out significantly from background extinctions, some of the lesser events that have been claimed may be open to question.

Those events that can, after rigorous analysis, be genuinely classed as catastrophic in a geological sense are unlikely ever to be pinned down in time more precisely than to a few tens of thousands of years, although exceptionally this limit may be reduced to a few thousand years. This conclusion is based on the study of particular examples in which ingenious juggling of inferred sedimentation rates in a given region is combined with the most refined radiometric dating that is possible.

Extinction scenarios that involve events as geologically instantaneous as a few months to a few years are consequently not amenable to rigorous testing from the stratigraphic record. We have to bear this in mind in what follows.

4

Impact by comets and asteroids

Although Norman Newell's pioneering research was published in 1967, general interest in mass extinctions provoked by catastrophic changes in the environment was not aroused until 1980, when a paper appeared in the journal *Science* proposing that the end-Cretaceous extinction was caused by the impact of a huge asteroid. Before this time several people had suggested an extra-terrestrial cause for particular mass extinctions. Thus, in the middle of the twentieth century, the German palaeontologist Otto Schindewolf, who had long been preoccupied with the marine mass extinction at the end of the Palaeozoic era, concluded on the evidence of fieldwork in the Salt Range of Pakistan that the event must have been a catastrophic one for which he could literally conceive no earthly explanation. He was consequently led to speculate that the causal factor was a nearby supernova explosion. The increased cosmic radiation impinging on the Earth could, he thought, have destroyed the ozone shield and have led to lethal exposure of numerous organisms. A few other such speculations invoking some kind of extraterrestrial factor were put forward at about the same time, and in 1970 Digby McLaren, an expatriate British palaeontologist who had risen to become Director of the Canadian Geological Survey, made a startling proposal. He was an expert on the late Devonian marine mass extinction at the end of the penultimate, Frasnian, stage. Like Schindewolf, he agreed that the event was much too wide-spread, dramatic, and 'geologically instantaneous' to have been caused by a merely terrestrial process, and he speculated that the world's ocean of the time had been severely disturbed by the impact of a giant meteorite. Three years later, the American chemist Harold Urey, a Nobel Prize-winner, pub-lished a paper in the journal *Nature* in which he argued that several extinction events within the past 50 million years had been caused by the impact of comets.

These various suggestions, together with a few others invok-ing increases in radiation from outer space, either in the form of cosmic radiation or solar protons, were virtually ignored. This is unsurprising in view of the almost total absence at that time of any supporting evidence, with the possible exception of a few tektite layers in Tertiary deposits. Tektites are glassy objects that were probably formed from terrestrial rock melted and displaced by the impact of a bolide. (*Bolide* is a convenient term that includes both asteroids and comets, since their effects on the Earth are not always readily distinguishable.) It was the tektite layers that had stimulated Urey's proposal. The situa-tion was to change dramatically with the publication in 1980 of the article in *Science*, an American journal with the same high prestige as *Nature*. Its authors, a group of scientists from the University of California at Berkeley, were the physicist and Nobel Laureate Luis Alvarez, his geologist son Walter, and the nuclear chemists Frank Asaro and Helen Michel.

The asteroid impact hypothesis

The origin of the impact hypothesis can be traced back to the beginning of the 1970s, when Walter Alvarez began a collaboration at the Lamont-Doherty Geological Observatory of Columbia University, New York with the Scottish geophysicist Bill Lowrie on the palaeomagnetism of deep-water pelagic limestones in Italy. Close to Gubbio in the Umbrian Apennines, a town with a beautiful medieval core, is an excellently exposed section in the Bottacione Gorge of a complete Cretaceous and Lower Tertiary succession. These limestones exhibit a very good record of magnetic reversals, which could be correlated with the biostratigraphic zonation based on planktonic foraminifera. The micropalaeontologists who worked on the section in the 1960s had discovered that there was a distinctive 1 cm clay bed exactly at the Cretaceous–Tertiary boundary, which was marked by the massive disappearance of nearly all the forams; only one of the smallest species continued into the Tertiary. Research in other parts of the world had shown that none of the species that disappeared reappeared in the Tertiary elsewhere, and so the event in question must mark a mass extinction. Such has been the interest in this thin clay, with numerous samples being taken for mineralogical and geochemical analysis, that its presence today is marked by a cleft in the section.

Alvarez was later reminded of something he had virtually forgotten: that the mass extinction had affected not only the planktonic forams (and the coccolithophorids, a group of nanoplankton that also had skeletons of calcium carbonate) but also such familiar groups as ammonites and dinosaurs. He was prompted to think that the boundary clay layer at Gubbio might possibly yield some evidence that had a bearing on the mysterious subject of extinction. In the summer of 1977 he therefore carefully collected some samples for analysis. When he joined the faculty at the University of California at Berkeley that autumn, he began to discuss the extinction problem with his father, Luis, who had been well established there as a particle physicist for many years and had recently retired. Luis's ever-curious mind was constantly on the lookout for new problems to tackle, and the long-standing enigma of dinosaur extinctions posed a great challenge. One obvious question to resolve was how long it had taken to deposit the clay layer. Luis Alvarez's suggestion was to measure its iridium content and that of the strata immediately above and below. Whereas iridium, which belongs to what is known as the platinum group of metals, is severely depleted in the Earth's crustal rocks, it is relatively enriched in meteorites, especially iron meteorites, though it is still present only in trace quantities expressed as parts per billion. Walter and Luis Alvarez assumed that iridium from meteoritic dust accumulated at a constant rate in sediments and thus made a useful geological timer.

Iridium had indeed already been used by geologists at the Scripps Institute of Oceanography in California to determine the cosmic content of Pacific floor sediments. The iridium content can be determined very precisely, to parts per trillion, by neutron activation analysis. Luis Alvarez's colleagues at the Lawrence Berkeley Laboratory, Frank Asaro and Helen Michel, were established experts with this technique, and were therefore invited in 1978 to collaborate in the research project. Incidentally, any of the platinum-group metals could be used for this purpose. It is just that iridium is easiest to analyse.

To everyone's surprise it turned out that there was far more iridium in the Gubbio boundary clay than could be accounted for by the normal rain of micrometeorites (**Fig. 4.1**). In the summer of 1979, while Walter Alvarez was in Italy doing field-work, his father set himself the task of finding an explanation for the huge iridium enrichment. For a while he toyed with the possibility that it could be explained by a nearby supernova explosion, but it turned out that the platinum-244 isotope that ought to characterize a supernova deposit was not present. So another hypothesis was required.

Further research had revealed that the iridium anomaly was essentially confined to the thin boundary clay, and that the iridium content quickly declined to normal background levels above and below in the stratal section. Comparable anomalies were found in Cretaceous–Tertiary (or K–T) bound-ary sections, as determined from fossil microplankton and nanoplankton (coccoliths) in both Denmark and New Zealand. This strongly suggested a global rather than a local event. Luis Alvarez eventually hit upon the idea that the Earth had been struck by an asteroid with the composition of a chondritic meteorite. (Chondritic meteorites are 'stony' meteorites that contain small globules called *chondrules*.) The impact of the asteroid would have expelled a huge amount of pulverized crustal rock, laced with iridium-rich asteroid material, to create a dust cloud that would have enveloped the Earth for up to a year. Such a cloud would have blocked out sunlight, thereby stopping photosynthesis and resulting in a collapse of the food chain, with consequent starvation and mass extinc-tion. The boundary clay at Gubbio and elsewhere marked

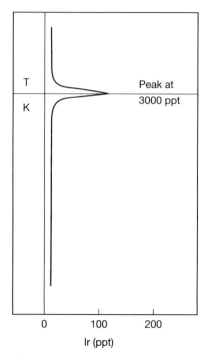

Fig. 4.1   Whole-rock iridium profile across 57 metres of
Maastrichtian–Palaeocene limestones at Gubbio, Italy, formed
in the open sea by the slow accumulation of sediments. The
beds depicted represent about 10 million years of deposition.
A pronounced iridium anomaly occurs in a clay layer 1
centimetre thick exactly at the Cretaceous–Tertiary (K–T)
boundary. This peak of 300 parts per trillion (ppt) of iridium
is flanked by 'tails' with iridium concentrations of 20–80 parts
per trillion that rise above the background level of 12–13 parts
per trillion. The fine structrure of these tails is the result of
diffusion and burrowing by organisms.
(Simplified from Alvarez *et al.* (1990).)

the material that had eventually settled from the globally distributed dust cloud. On this assumption, and knowing the chemical composition of chondritic meteorites, Alvarez calculated an approximate diameter for the asteroid of 10 km, which would imply the creation of an immense impact crater some 150–200 km in diameter.

Early reactions to the hypothesis

The 1980 paper in *Science* quickly provoked great excitement among a variety of scientists, including astronomers, physicists, and chemists, but palaeontologists, with a few exceptions such as Digby McLaren, were decidedly sceptical, if not entirely dismissive. There was no evidence that terrestrial and marine extinctions were strictly contemporary. Too little, it was claimed, was known about the geochemistry of iridium. The iridium could have been concentrated by organisms in the sea, or selectively adsorbed by organic matter on the sea floor. Better stratigraphic control was needed, because iridium anomalies might prove to be common in the geological column once they were looked for, and not always associated with mass extinction events. Where was the huge crater demanded by the hypothesis? The key argument, however, was that, with the exception of the planktonic foraminifera and the coccolithophorids, whose extinction record did seem genuinely catastrophic, those Cretaceous organisms that had become extinct at the end of the period did so with a whimper rather than a bang. This was evidently the case with many familiar fossil groups that seemed particularly characteristic of the Cretaceous, and sometimes of the Mesozoic as a whole. These include ammonite and belemnite cephalopods, inoceramid and rudist bivalves, and ichthyosaur and pleiosaur marine reptiles.

Consider the case of the dinosaurs, the only major land group to disappear at the K–T boundary. Although there were exceptions, such as the Canadian Dale Russell, the consensus of vertebrate palaeontologists was that the dinosaurs were already on their way out, for thousands or even millions of years beforehand. In order to believe that the asteroid impact was wholly responsible for their ultimate demise, one would be obliged to indulge in special pleading and maintain that the dinosaurs were clairvoyant and died of fright. The reaction of Luis Alvarez to all this was characteristically irascible. He dismissed palaeontologists as being not proper scientists but mere stamp collectors. This comment recalls Ernest Rutherford's remark early in the twentieth century. When asked about what constituted science, he replied that there was physics, chemistry, which is a sort of physics, and stamp collecting. Thus the whole of biology as well as geology was airily dismissed, to say nothing of the social sciences. If one replies that today Rutherford would, perhaps grudgingly, concede a place at high table for molecular biology, what about the type of research that Darwin engaged in? Is there no place for keenness of observation, depth of insight, imaginative interpretation, and capacity for synthesis, in all of which Darwin excelled? Fortunately, the collaborative research across a wide range of disciplines that mass-extinction studies have generated in the past two decades has seen little of this traditional arrogance of physicists towards other sciences.

Later developments

The Berkeley group's findings stimulated a great deal of geo-chemical and mineralogical research, and before long many other K–T boundary horizons in widely distributed marine deposits revealed an iridium anomaly. The ratio of other siderophile (or 'iron-loving') elements, such as platinum and osmium, and related elements, such as gold, were found to be enriched in some key localities such as Stevns Klint in Denmark. Another important discovery was made by Carl Orth and his American colleagues. They found an iridium anomaly in a non-marine section in New Mexico, precisely at the K–T boundary as defined by pollen grains. This confirmed a prediction of the impact hypothesis, that the anomaly should be global, and appeared to rule out decisively an alternative hypothesis that the iridium was extracted from sea water. Not long afterwards the Berkeley group found another iridium anomaly in a non-marine K–T boundary section in Montana. Furthermore, palynologists and marine micropalaeontologists established to their satisfaction that the K–T boundary in the pollen record was contemporary with that determined by foram workers, and both continental and marine boundaries were found to occur in the same magnetic zone, 29R. (Rever-sals of the Earth's magnetic field in the geological past make it possible to establish a geomagnetic polarity timescale that can be used to correlate rocks in which a primary magnetization can be measured.) The Dutch micropalaeontologist Jan Smit was the first to recognize crystalline spherules in the boundary clay in southern Spain. Similar spherules have subsequently been found elsewhere, and have been interpreted by the impact supporters as cooled droplets of impact melt – in other words microtektites. When it was recognized that the eruption of a huge flood-basalt province in India known as the Deccan Traps embraced the K–T boundary (see Chapter 8), a group of scientists challenged the impact theory in favour of a volcanic explanation, arguing that the spherules as well as the iridium anomaly were volcanic in origin. Two other developments, however, greatly strengthened the impact theory.

The first of these developments was the discovery of what is termed 'shocked' quartz in the K–T boundary clays in the Western Interior of the United States and Canada and, less convincingly, in other parts of the world. Quartz grains as seen under the microscope in thin section normally display no planar features because, unlike many other minerals, they lack even cleavage. Shocked quartz, however, displays multiple sets of planar features known as shock lamellae, which are thought to be produced only by the imposition of huge pressures such as occur at nuclear explosion sites or meteorite craters (**Fig. 4.2**). Shocked quartz was first found by Bruce Bohor of the US Geological Survey at the K–T boundary in Montana and reported in a paper in 1984. Comparable material was quickly found elsewhere in the Western Interior. The mid-1980s were dominated by debate between the 'impactors' on the one hand and the 'volcanists', led by Charles ('Chuck') Officer at Dartmouth College, New Hampshire, on the other. This debate came to a head at a major international conference at Snowbird, Utah in 1988. The clear impression gained by most of those who were there was that the impactors won on the subject of shocked quartz, which does not seem to be produced by any amount of volcanism, because the pressures generated by volcanic activity are never high enough.

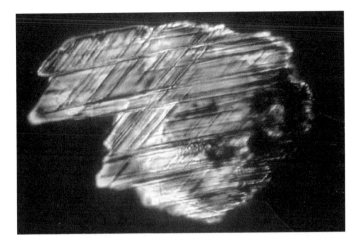

Fig. 4.2   An example of shocked quartz, displaying the characteristic multiple sets of planar features known as shock lamellae. G. A. Izett/US Geological Survey.

Was there yet more independent evidence to be found? The obvious 'smoking gun' would be a plausible, enormous impact crater of the right age. The lack of such a crater had been a major point of criticism by the sceptics. Claims of a deep-sea location lost from recognition today because of subsequent sedimentation became implausible because of the shocked quartz, which suggested a continental rather than an oceanic source. (The ubiquitous basalt that forms the ocean floor beneath young sediments lacks quartz.) The size and abundance of the shocked quartz in relation to samples found elsewhere suggested a site somewhere on the North American continent. Eventually, as the result of a geophysical survey in conjunction with evidence from boreholes, such a site was

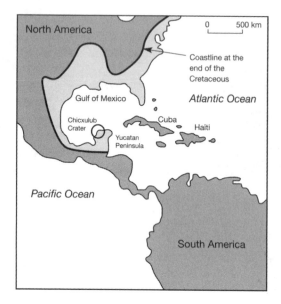

Fig. 4.3   Map of the Gulf of Mexico–Caribbean region showing the location of the purported Chicxulub crater and the approximate location of the coastline at the end of the Cretaceous period. (From A. Hallam and P. Wignall, *Mass extinctions and their aftermath*, Oxford University Press, 1997.)

found at Chicxulub, on the margin of the Yucatán Peninsula in Mexico (**Fig. 4.3**). By the early 1990s pretty well all the Earth sciences community, including palaeontologists, had been persuaded of the reality of a huge impact. The only important dispute then concerned the size of the impact crater: was it 170 km or 300 km in diameter? Either would suggest a huge impact by either an asteroid or comet. Gene Shoemaker, the leading planetary geologist of his day, favoured a comet because of the much greater energy of impact. However, because comets are thought by astronomers to be no more than 'dirty snowballs' it is hard to see where all the iridium in the K–T boundary beds would have come from.

Possible environmental scenarios

The environmental effects invoked by supporters of bolide impact include severe reductions in light levels, changes of temperature, acid rain, and wildfires. In the original scenario set out by Alvarez and his colleagues, a world-embracing dust cloud, composed of expelled crustal rock laced with meteoritic material, severely reduced light levels, thereby inhibiting photosynthesis and hence adversely affecting the base of the food chain, both on the continents and in the oceans. Theoretical and experimental evidence suggests that significantly reduced light levels maintained for as little as a few days would eliminate 99 per cent of the zooplankton. These may be underestimates if massive wildfires are also invoked, because soot absorbs sunlight more effectively than rock dust. So far as short-term temperature change is concerned, an early suggestion was that water vapour in the atmosphere would induce a global greenhouse effect. The dominance of calcium carbonate and calcium sulphate minerals in the upper 3 km of the Yucatán section has led to active research on the effects of devolatilization. Experiments have indicated that expulsion of CO_2 from the limestone target could have given rise to a global warming of 2–10 °C. On the other hand, sulphate aerosols could have led to several years' cooling of the atmosphere. Other models have indicated that significant cooling could have lasted for a few years, followed by a more prolonged period of moderate warming, because of the longer residence time of CO_2 in the atmosphere. Such a timescale is unfortunately far too short to leave an oxygen isotope record in the strata, from which trustworthy temperatures could be inferred.

Impact-induced acid rain on a massive scale has also been invoked to account for the destruction of calcareous plankton (foraminifera and coccoliths) by making the surface waters more acid. Doubt has, however, been expressed that the chemistry of the global surface ocean would have been significantly affected, because of the strong 'buffering' of sea water; and modern plankton culturing studies are not readily compatible with scenarios that rely on acid rain as the primary cause of extinction. There is also a problem with the high survival rate of freshwater vertebrates such as crocodiles, because modern lakes suffering from acid rain are substantially stripped of life.

An analysis of soot concentration in the purported impact fallout boundary clay layers in Denmark, Spain, and New Zealand by Wendy Wolbach and her colleagues revealed a striking enrichment not hitherto found anywhere else, and this finding led to the hypothesis of a global wildfire induced by impact. This initial study was open to the criticism that it assumed that all the clay was deposited within the geological instant associated with impact. If that is not the case, and the clay is a normal one, locally derived, then such a concentration is not significantly different from those found in the sediments stratigraphically above and below the clay. Wildfires on at least a modest scale were no doubt commonplace in geological history, and the mere occurrence of soot is of no great significance; the crucial point is whether the amount of enrichment is unusually large. Subsequent work by Wolbach and her colleagues made some allowance for this problem, but global wildfires were still insisted upon. The wildfire hypothesis nevertheless suffers from several difficulties. In the first place, so little work has been done using a rather specialized technique that there is almost a total absence of control, in marked contrast to the situation with iridium anomalies. In effect, hardly anything is known about the concentration of soot in the stratigraphic record as a whole. Furthermore, it is surprising that no abnormal concentrations of soot have been found, despite intensive search, in one of the most obvious places to look, the North American Western Interior. Finally, the effects of the claimed global wildfires would have been far too drastic to account for the large proportion of survivors in the continental record.

It is time to consider in more detail what the fossil record across the K–T boundary, continental and marine, is now thought to tell us, but some further geochemical information should first be outlined.

In the present ocean system the calcareous skeletons of plankton are characterized by positive carbon-13 values because the dissolved bicarbonate in surface waters is relatively depleted in carbon-12 owing to photosynthesis. On the other hand, bottom waters are relatively enriched in carbon-12 as a result of respiration of isotopically light organic carbon, and benthic (bottom-living) foraminifera are characterized by above-average carbon-12 values. For this reason a positive surface- to bottom-water gradient develops. If for some reason there is a drastic decline in organic productivity in surface waters, the gradient disappears, and carbon-13 carbonate will record a strong negative shift. A phenomenon of this type is recognized very widely in the record of oceanic borehole cores that penetrate the K–T boundary. This fact led Ken Hsu, the cosmopolitan geologist born in China, who had a long spell in the United States and is now a Swiss citizen, to propose the term 'Strangelove Ocean' for this condition, which he named after the mad scientist memorably played by Peter Sellers in the eponymous film directed by Stanley Kubrik. The kind of catastrophe explored with excessive enthusiasm by Dr Strangelove was a nuclear holocaust, but the mass destruction of life, leading to a fall in plankton productivity, could have another cause, such as an extraterrestrial impact.

Biotic changes across the Cretaceous–Tertiary boundary

The exciting discoveries made in geochemistry, mineralogy, and geophysics have persuaded the great majority of palaeontologists, together with the majority of other Earth scientists, of the reality of a major impact event in Mexico. Argument nevertheless persists about the significance of this impact for the extinctions at the K–T boundary. For this, the only evidence can come from the fossil record. Is this record indeed consistent with a catastrophic event, as required by the impact hypothesis? The great majority of palaeontologists with expert knowledge of the fossil groups involved have in the past argued for gradual rather than catastrophic change. Impact enthusiasts such as Dave Raup, Digby McLaren, and Ken Hsu have scornfully dismissed these people as having been indoctrinated, like virtually all geologists trained earlier in the twentieth century, by Lyellian gradualism. So what does the record actually show?

The answer is not an easy one to arrive at. In the first place there are numerous fossil groups to consider, and the study of each group requires considerable technical knowledge that is available only to those experts who have made their professional careers on the groups in question. Thus it is not just those scientists who are not palaeontologists, but palaeontologists who study other groups who are obliged to take the opinions of these experts on trust. Secondly, the number of stratigraphic sections with an adequate record is decidedly limited, and few have been subjected to the kind of intensive statistical and cladistic testing required to eliminate problems such as the Signor–Lipps effect or pseudoextinction, mentioned in Chapter 3.

Let us first consider the all-too-familiar dinosaurs, which unfortunately have a relatively poor fossil record compared with most marine organisms. Indeed, the only really good stratigraphic sections that record a continuous non-marine succession across the K–T boundary are in the North American Western Interior, most especially in eastern Montana. Elsewhere in the world the dinosaurs could have died, for all we know, before the end of the Cretaceous or survived into the Palaeocene, as has been claimed by some palaeontologists working in China. The sections in eastern Montana have provoked a sharp disgreement between the 'gradualists' and the 'catastrophists'. In an attempt to resolve this dispute, Peter Sheehan of the Milwaukee Public Museum in Wisconsin enlisted the help of a large number of volunteers over several years to scour the hillsides for dinosaur bones. These volunteers put in an impressive number of man-hours in the process. After expert analysis of what they found, the conclusion was drawn that there was no decline in diversity up the stratigraphic succession until the very end of the Cretaceous, when the dinosaurs dramatically disappeared. Despite some further argument with gradualists such as Dave Archibald, it is difficult to refute this impressively obtained result – but that does not mean that the dispute in a broader context is over. Thus, a significant decline in generic diversity in the Western Interior over several million years has been claimed from the late Campanian, the penultimate stage of the Cretaceous, with 45 genera, to the Late Maastrichtian, the last stage, with 24 genera. This strongly suggests that the dinosaurs might already have been on their way out when an impact delivered the *coup de grâce*.

Other vertebrates, both on land and in the sea, do not have a much better record. What evidence we do have tells a very varied picture. On land the marsupial mammals were more severely affected than the placentals, but again there is little good evidence outside western North America. On the other hand, crocodiles and freshwater turtles survived the K–T event virtually unscathed. Antique kinds of birds seemed to have disappeared, with modern types radiating in the early Tertiary, but the fossil record is very poor and there is a dispute between palaeontologists and molecular biologists: the latter argue on the basis of their phylogenetic reconstructions, that many modern bird orders originated well back in Cretaceous time. The other major land group of animals for which we have a better than expected record are the insects. Unlike the events at the end of the Permian, when this group suffered a major extinction, there appears to have been no significant change across the K–T boundary. Returning to vertebrates, this time to sea-living creatures, fish do not seem to have been greatly perturbed. Pterosaurs no doubt flew over land as well as sea, but their only good record is in marine deposits. They became extinct at the end of the Cretaceous, as did the swimming reptiles that had survived for most of the Mesozoic, the ichthyosaurs and plesiosaurs. Here again, the record is not good enough to pronounce on how catastrophic their demise was, but the ichthyosaurs and plesiosaurs had been in long decline and could well have completely disappeared before the end of the Cretaceous. The most promising vertebrates for those who favour a catastrophic end-Cretaceous event are a group of giant swimming reptiles, the mosasaurs, which arose and radiated rapidly in the late Cretaceous, only to disappear as dramatically by the end of that period. The record for other taxa, however, is inconclusive as to whether or not the disappearance of the mosasaurs coincides precisely with an impact event.

By far the best record to establish true catastrophism is that of those microscopic fossils derived from plankton that can occur in myriad quantities in small volumes of rock, the coccoliths and foraminifera. In the fine-grained marine sediments in which these fossils occur in greatest abundance, the record is unequivocally in favour of some sort of dramatic turnover of species within a very short stratigraphic distance across the K–T boundary. Whether the catastrophic extinction took place over a year or two at most, as required by the impact hypothesis, or over several tens of thousands of years, remains a matter of dispute. There has been little dispute about the coccoliths, for which the turnover of almost all species appears to be consistent with an Alvarez-type scenario, but the situation is quite different for the planktonic foraminifera. At one extreme, Jan Smit of the University of Amsterdam has argued for an almost complete turnover of species at the K–T boundary, which is entirely consistent with the impact hypothesis. At the other extreme, Gerta Keller of Princeton University maintains that the extinctions began up to a few tens of thousands of years before the end of the Cretaceous, and that there was a higher proportion of survivors into the early Palaeocene than is claimed by Smit. In conjunction with her former post-doctoral associate Norman MacLeod, now at the Natural History Museum in London, Keller has endeavoured to show that many of the K–T boundary sections studied in borehole cores are unusually condensed because of a sharp rise of sea level, thereby giving a misleading impression of the rate of species turnover. Furthermore, biogeographical work by Keller and MacLeod has indicated that the catastrophic change was confined to tropical and subtropical regions, and was not a worldwide phenomenon, as one might expect from the Alvarez hypothesis.

It is difficult for outsiders to assess critically the respective merits of these opposing views, which depend not just on careful collecting but the quality of the taxonomic work used to distinguish the various species. It would be fair to say that most experts have tended to side with Smit, but on the other hand a 'blind-test' analysis of samples from the global type section of the K–T boundary at El Kef, Tunisia, proved somewhat indecisive in choosing between the Smit and Keller interpretations.

There is no need to undertake a comprehensive review of what the marine invertebrate record appears to tell us, but several points are worth mentioning. Some important Mesozoic groups such as the ammonites were in slow decline throughout the latter part of the Cretaceous, but Peter Ward's careful collecting in the Basque country has established that at least in that region there was a dramatically sudden disappearance of the remaining species. Among the bivalves, two other important groups of molluscs, the rudists and inoceramids, both of which are common and characteristic in the Cretaceous, became extinct in the Maastrichtian, perhaps as much as a million years before the end of the period. Other bivalves appear to have undergone a catastrophic extinction at the end of the Cretaceous, as some thorough statistical analyses have shown. Other invertebrate groups exhibit a range of patterns, from increasing rates of extinction through the Maastrichtian to catastrophic extinction at the end. 'Catastrophic' has here to be interpreted in the sense that geologists understand catastrophes, as events affecting a large proportion of a fauna within thousands rather than millions of years. Many of the groups that evidently suffered catastrophic extinction were organisms higher in the food chain than the plankton, and so their extinction could be seen as being a mere consequence of plankton extinctions.

Among the terrestrial plants there was a dramatic increase in extinction rate at the K–T boundary in western North America, where there is a good terrestrial stratal record, which coincides with a so-called 'fern spike', recognized by palynologists, which marks the expansion of fern populations after disaster had struck the 'higher' plants, namely the angio-sperms. This appears to be consistent with the impact scenario, but the event seems to have been limited to North America. Elsewhere in the world, and particularly in the southern hemi-sphere, no more than gradual changes across the boundary can be perceived. More generally, the terrestrial record (based on pollen and spores) and the marine record in Seymour Island, Antarctica, which is the best K–T boundary section in the southern hemisphere, shows gradual rather than catas-trophic change.

One of the most interesting attempts to explain selective extinctions at the boundary has been made by the American palaeontologists Peter Sheehan and Thor Hansen. In the marine realm, filter-feeding invertebrates living on the sea bed would have been very vulnerable to disturbance of their plank-tonic food supply, and all these groups experienced sub-stantial extinctions. On the other hand, detritus feeders, carnivores, and scavengers would have been relatively resist-ant and apparently experienced comparatively low extinction rates. Sheehan and Hansen also applied their model of detritus feeding as an extinction buffer to continental environments. Today's niche of mammalian browsers was occupied in the late Cretaceous by herbivorous dinosaurs, on which the minority of carnivorous dinosaurs would have been dependent for food. Damage to the plant world by inhibition of photosynthe-sis, as envisaged in the Alvarez scenario, would inevitably have had adverse consequences for the dinosaurs. On the other hand, many, if not most, of the small mammals would have been insectivorous. The high survival rate of turtles, croco-diles, and champsosaurs can be explained by the fact that freshwater ecosystems are based on land-derived detritus.

The biological story is clearly very complex, much more so indeed than the brief summary above indicates. The fairest generalization we can make from the evidence so far indicates increasing extinction rates for many groups within the Maastrichtian, and that the end-Cretaceous impact event could have served as no more than a final blow to an already bruised and battered biosphere.

Is there a relationship between impacts and other mass-extinction horizons?

After the success of the impact research agenda for the K–T boundary, great enthusiasm was expressed by some scientists that they had perhaps found the key, not just to the K–T event, but to all mass extinctions. This was certainly Luis Alvarez's hope, if not expectation, before he died, and the view of at least one distinguished palaeontologist, Dave Raup. In fact bolide impact was seized upon by many neo-catastrophists as a kind of deus ex machina. This Latin phrase, usually translated as 'god out of the machine', actually derives from the ancient Greek theatre. *Machina* means 'scaffolding' as well as 'machine', and in the appropriate context 'god out of the scaffolding' is the better translation, because it refers to actors playing gods descending to the stage. Thus a writer finding difficulties in ending his play or his novel, might resort to some dramatic and perhaps implausible resolution. In the context of mass extinctions, introducing a deus ex machina implies that there is no terrestrially bound process that is adequate to do the job: in other words there is a 'plot failure'. Once again, the only proper scientific approach is to seek supporting evidence.

Two approaches can be adopted. One is to seek evidence for impact, such as iridium anomalies, shocked quartz, tektites, and impact craters, at established mass-extinction horizons. The alternative is to examine the stratigraphic record for evidence of this kind and see to what extent it correlates with mass-extinction horizons.

We have seen that the K–T event scores well as a horizon recording a time when a major impact took place, but what about the other four of the 'big five', or for that matter a larger number of lesser events? Most attention has, not surprisingly, been devoted to the biggest event of all, which occurred at the end of the Permian. An early claim was made of an iridium anomaly in the classic Meishan section near Nanjing, subsequently accepted as the global type section, but this was not confirmed by two distinguished American nuclear chemists, Frank Asaro, a member of the original Alvarez team, and Carl Orth. Furthermore, no convincing evidence of shocked quartz, tektites, or craters at this horizon has been put forward from anywhere in the world. This has not stopped persistent attempts to provide other evidence. For example, a group of American geochemists has examined further material from Meishan in the form of carbon known as fullerenes. They have claimed that the helium and argon isotope ratios of these fullerenes resemble those of the meteorites known as carbonaceous chondrites, and have put this forward as evidence of impact. However, other geochemists have challenged these results, which even if substantiated could signify no more than that, because the Meishan section is so highly condensed, the isotope ratios could merely be recording in the sediments the increased proportion of micrometeorite background 'rain'.

A group of Japanese geochemists has recognized a significant shift across the Permian–Triassic (P–T) boundary at Meishan towards a lower sulphur isotope ratio. In conjunction with an increase in the amount of nickel at one horizon they interpreted this as a consequence of a bolide impact that caused substantial submarine volcanism. This interpretation has, however, been strongly attacked by a group of impact experts from across the world, led by Christian Koeberl of the University of Vienna. They conclude that none of the points raised by the Japanese workers, Kaiho and colleagues

> provides even a vague suggestion of an impact event at the P–T boundary. While an impact event is one of several possibilities to explain this mass extinction, the interpretations presented by Kaiho *et al.* are poorly documented, inconclusive, and bypass more obvious explanations of the data. Attempts to utilize the questionable interpretations by Kaiho *et al.* to support the equally controversial . . . claims for the presence of extra-terrestrial ^{3}He in fullerenes at the P–T boundary represent circular logic.

In the normally restrained language of scientific discourse this dismissal could hardly be more unequivocal.

What I found most intriguing of all is the work of a group of Chinese palaeontologists on the same famous section at Meishan. Their intensive statistical study of the stratigraphic distribution of a large number of fossils purported to show that the mass extinction across the P–T boundary was indeed catastrophic, and thus compatible with a bolide impact. I accepted an invitation to referee their paper, which had been submitted to *Science*. While I found their statistical results persuasive in favour of a catastrophic event, I thought their conclusion highly tendentious in view of the fact that they had received financial support from a Chinese Astrogeology organization. This is the kind of thing one is perhaps more likely to encounter in some of the more dubious pharmacological research, which supports the claims made for a particular drug, and which is financed by the drug company that manufactured it. My recommendation to the editors of *Science* was that the paper should be published because of the good quality of the palaeontological data but that the concluding sentence referring to bolide impact was unjustified by the evidence presented and should be deleted. In the event, the authors retained the offending sentence and deleted instead in their acknowledgement all reference to their paymasters!

We now turn to the Triassic–Jurassic (T–J) boundary. An intensive study of a boundary clay in Austria failed to reveal evidence of either an iridium anomaly or, as had been claimed by a Russian geologist, shocked quartz, which he indeed found at several horizons well below the boundary. An inspection of quartz grains in the boundary clay by Richard Grieve, one of the leading impact experts in North America, failed to confirm this claim, and the rather curved laminae found in a few of the grains were dismissed by Grieve as being caused by tectonic forces of the sort produced in mountain-building. Another claim for shocked quartz, made for a purported T–J boundary section in the northern Apennines, has been much quoted in support of an impact event at this horizon. There are problems with this interpretation. In the first place there is no biostratigraphic proof that the T–J boundary has been accurately located in this section. Secondly, the shocked quartz grains, if that is what they are, are rare and occur in at least three horizons. To explain this the authors interpreted this result to indicate impact by a shower of comets over a geologically extended period of time. Unfortunately the idea of comet showers, introduced for the K–T boundary and the late Eocene by the astronomers Piet Hut, Gene Shoemaker, and a number of palaeontologists to account for extinctions over an extended period of time, had subsequently been abandoned. This is because there is evidence of only one impact event at the end of the Cretaceous, and because no discrete mass-extinction horizon is recognized in the late Eocene. Furthermore, Alessandro Montanari, Italy's greatest impact enthusiast and expert, who had introduced the Americans to the section in question, was unable to confirm their results from his own subsequent work. It seems most probable that, as in Austria, the presumed shocked quartz has been produced by terrestrially bound tectonic forces.

More recently a considerable stir was caused by an article in *Science* by Paul Olsen and his colleagues, published in 2002. Olsen, who researches at the Lamont-Doherty Geological Observatory in New York State, is the world's leading authority on the 'Newark Supergroup', a succession of non-marine strata in eastern North America straddling the Triassic–Jurassic boundary. He has also long been an enthusiastic advocate of the impact theory. Despite the lack of marine fossils Olsen believes that he can fix the boundary on the basis of pollen. He reported in this paper an iridium anomaly at the purported boundary which coincides with a change in the type of vertebrate footprints found, suggesting an early phase of dinosaur extinction. Unfortunately for the impact hypothesis, the anomaly is far too small to provide decisive evidence for impact, and this was pointed out by Richard Kerr in a commentary in the same issue of *Science*. There are plenty of terrestrial causes that could account for such a modest anomaly. Furthermore, an intensive mineralogical study of strata at the purported T–J boundary within the Newark Supergroup has failed to find any shocked quartz. The evidence from the footprints is interesting, but it needs to be confirmed by other vertebrate specialists, who have hitherto been sceptical about any significant faunal change across the boundary.

In the wake of McLaren's giant meteorite hypothesis to account for the Frasnian–Famennian (late Devonian) extinction event, there has been an active search across the world for supporting evidence. No convincing iridium anomaly has been found, and shocked quartz has not even been claimed. Horizons of microtektites have indeed been found in both Belgium and China, but these do not coincide with extinction events. So far as the end-Ordovician mass extinction is concerned, there is a decided lack of evidence for impact, and no serious claims have been made. Sporadic attempts have been made to claim correlations with lesser extinction events, such as at the Cenomanian–Turonian boundary (mid-Cretaceous) but have proved unconvincing to the majority of Earth scientists.

Let us turn now to the second approach, which entails seeking convincing evidence for impact independent of mass extinctions. On some occasions small iridium anomalies have been reported, but such anomalies can also have a terrestrial origin; for example, extreme condensation of the containing strata can be the result of an exceptionally low rate of sedimentation, so that micrometeorite dust is concentrated. The best record of convincing tektites or microtektites comes from the Palaeogene (early Tertiary). Upper Eocene impact-generated microtektite layers are widely distributed in the low-latitude regions of the Pacific, Indian, and Atlantic oceans. Considerable debate has taken place about how many bolide impacts can be inferred for the late Eocene, the number ranging from one to six. The number depends on the best bio-stratigraphic estimate based on microfossils, and one impact seems more plausible than several. The most probable site of the impact crater, which is associated with a tektite layer traced across the south-eastern United States and Caribbean, is in Chesapeake Bay. For none of the tektite layers found in different regions is there any association with mass extinction of the microfauna found in the borehole cores containing the tektites. Montanari and his colleagues have found iridium anomalies of late Eocene age both in Italy and from a borehole core on the Maud Rise, off Antarctica, which can plausibly be related to the impact event or events.

The Earth's cratering record also provides pertinent evidence for impact, and many well-authenticated Phanerozoic craters are now recognized. This information can be used to test a model, namely Raup's bleakly termed 'kill curve'. Current estimates of Phanerozoic impact rates are combined with the kill curve to produce an impact-kill curve, which predicts extinction levels from the diameter of the crater. Raup's model can be applied to some of the best-authenticated impact craters.

To begin with, let us take the Ries Crater of Bavaria, which has a diameter of about 30 km and was formed about 15 million years ago, in the Miocene epoch. Close to the middle of this crater, the rim of which forms a line of hills, is the pretty little town of Nördlingen, which has an excellent museum in its centre devoted to explaining the crater. The characteristic rock produced by the impact is composed of large fragments of what geologists call a breccia, but this particular example is known as suevite. It is exposed on the crater margins in a few quarries but can be most conveniently seen as the building stone of the principal church. According to the Raup curve about 10 per cent of species should have been made extinct for a crater of this size, but no species extinctions are recognizable from the region for either mammals or plants, the two fossil groups that have been studied in the Miocene strata. In other words, while the local area must no doubt have been like a moon landscape for a considerable time on a human timescale, all the species of the region evidently returned rather quickly by geological standards.

The Montagnais impact crater on the Nova Scotia shelf is 45 km wide and is dated as late early Eocene in age. The Raup curve indicates a 17 per cent species extinction, but there is in fact no recognizable biological change on a local, regional, or global scale. The three largest craters in the Phanerozoic record, apart from Chicxulub, which is the feature invoked for the K–T boundary extinctions, are the Popigai in Siberia, the Manicouagan in Quebec, both about 100 km in diameter, and the Morokweng in South Africa. The Popigai crater is dated as about 39 million years old, and Raup has claimed that it could have been associated with end-Eocene extinctions. Both marine and continental extinctions at this time extended, however, over several million years and there is no clear-cut end-Eocene event. Furthermore, 39 million years is several million years before the Eocene–Oligocene boundary. The most recent dating of the Manicouagan crater is about 214 million years, about 14 million years before the end of the Triassic, corresponding to the early Norian stage, a time for which no one has recognized any extinctions, either in the marine or continental realm. Yet, according to the Raup curve, no fewer than 50 per cent of species should have become extinct. The Morokweng crater, discovered more recently, has a crater diameter of at least 70 km, and possibly much more. It is dated as at about 145 million years, which corresponds approximately to the end of the Jurassic. Although a minor mass extinction for this time has been claimed by Raup and Sepkoski (see Chapter 9), it has not been recognized by other scientists.

We are obliged to conclude from a more general study of the Earth's cratering record, as determined from a combination of gravity and aeromagnetic surveys, shocked quartz, and the analysis of impact melt rocks, that, contrary to Raup's opinion, bolide impact cannot plausibly be invoked as a general cause of mass extinctions. Even for the one example where a good case can be made for coincidence with a significant impact, at the end of the Cretaceous, it can be seen that the extinction story is more complicated than that originally proposed by the Alvarez group, and that other terrestrial factors must be invoked. So it is now time to go on to consider what at least some of these factors might have been.

5

Sea-level changes

In earlier times many geologists clearly became cynical about what they had learned as students about Earth history from their stratigraphy courses. 'The sea comes in, the sea goes out.' This is indeed one of the most striking signals that emerges from study of the stratigraphic record in a given region: a succession of marine transgressions and regressions on the continents. Little scientific rigour was, however, applied to the subject, and students were left with no overarching explanation to provide any intellectual stimulation. Since the 1970s things have begun to change for the better, as less emphasis has been placed on learning the names of rock formations and fossil zones and more on the dynamic aspects of what to many ranks as a fascinating subject. This entails studying changing geographies and climates within a framework supplied by plate tectonics, the successions of strata being subjected to ever-more-rigorous sedimentological and geochemical analysis, and global correlation continually improved by further study of stratigraphically useful fossils.

How do we infer sea-level changes from a given succession

of sedimentary rocks? In essence we use *facies analysis*, which is based upon a careful study of the sediment types and structures, together with a study of the ecological aspects of the contained fossils, or *palaeoecology*. These features can be compared with those of similar sediments that are being deposited today, or similar organisms living today. Comparisons of this kind were practised by the likes of Cuvier as well as Lyell. Consider, for example, the Cretaceous succession in southern England. The oldest strata, well exposed on the coast from Sussex to Dorset (**Fig. 5.1**), are known as the Wealden, and consist mainly of sandstones and siltstones that were deposited in a coastal plain (non-marine) setting. They are overlain by the Lower Greensand, a sandy unit of Aptian–Lower Albian age laid down in a very shallow marine environment. These conditions are revealed, not just by the types of fossils, which include the exclusively marine ammonites, but also by the distinctive green clay mineral glauconite, which gives its name to the rock formation and occurs today only in marine settings. The unequivocal implication is that the sea level rose in Aptian times.

The trend continues into the younger Cretaceous, because the Lower Greensand is overlain first by the Gault Clay, a finer-grained deposit of Middle and Upper Albian age, laid down in deeper water, and then by the even deeper-water Chalk, of Cenomanian and younger age. Chalk is an unusual type of limestone composed largely of the remains of calcareous microplankton (coccoliths and foraminifera) which becomes steadily purer and more devoid of terrestrially derived sediments as one follows it up the succession. In other words, the Cretaceous exhibits a 'deepening-up' succession. It follows logically that there should be a corresponding

Fig. 5.1 Photograph looking eastwards towards Lulworth Cove,
Dorset, a county where the magnificent Jurassic and
Cretaceous cliffs have been designated a World Heritage Site.
The rocks forming the headland consist of alternating
limestones and shales of the Lulworth Formation straddling
the Jurassic–Cretaceous boundary, which were deposited in a
marginal marine-to-lagoonal environment. (These sediments
are traditionally known as part of the Purbeck Beds.) They are
overlain by soft-weathering silts, sands, and clays of the early
Cretaceous Wealden Beds, also non-marine, into which the
cove has been excavated by the sea. They pass up into shallow
marine greensands and then into deeper-water marine Chalk,
signifying a more or less progressive rise of sea level from the
early to the mid-Cretaceous. Stair Hole in the foreground
shows tectonically disturbed Purbeck Beds, known as the
Lulworth Crumple.

spread of marine deposits, marking a transgression of the sea (**Fig. 5.2**). This is in fact clearly demonstrated, first in midland and northern England, where the non-marine Wealden strata are absent and are unlikely ever to have been deposited, and the oldest Cretaceous is the Lower Greensand. The second clear instance of transgression is in north-western Britain (Northern Ireland and the Scottish Hebrides), where Chalk rests directly upon much older Jurassic rocks, there being no evidence whatever that any older Cretaceous rocks were ever deposited there. It is reasonable, therefore, to infer the existence of land in this region directly before the deposition of the Chalk.

Such changes are often of no more than limited geographic extent and are due to regional subsidence of the substratum. Correspondingly, evidence of regression giving rise to a shallowing-upward succession is due to regional uplift or to the seaward advance (prograding) of a river delta. Less frequently, however, biostratigraphic correlation can demonstrate that the changes are not merely regional but global in extent. Sea-level changes of this sort were termed *eustatic* in the late nineteenth century by the great Austrian geologist Eduard Suess. One of his best examples is what came to be widely known as the Cenomanian transgression, which made what was evidently a spectacular mid-Cretaceous inundation of the continents. What can be observed in Great Britain is indeed the local expression of a truly global event of major importance.

Suess's proposal of sea-level changes of global extent quickly met with a positive reception among geologists studying the Quaternary, because of the clear evidence that began to emerge that the polar ice caps had expanded and retreated on

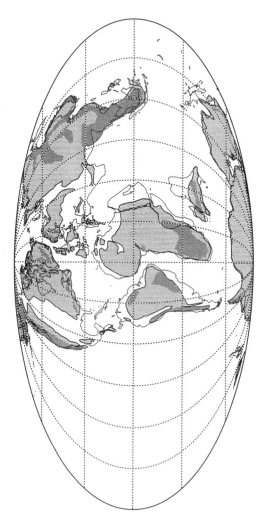

Fig. 5.2 The global extent of the mid-Cretaceous marine transgression, about 100 million years ago. Stippled areas indicate land. (After Alan G. Smith, David G. Smith, and Brian M. Funnell, *Atlas of Mesozoic and Cenozoic coastlines*, Cambridge University Press, 1994, p. 38.)

a number of occasions during that time in response to global changes of climate. In cooler times polar ice grew in extent at the expense of sea water, and in warmer times melted ice was restored to the ocean. Thus, glacial episodes produced regressions of the sea, sometimes to the edge of the continental shelf, whereas interglacials produced transgressions. The speed of these changes in the Quaternary, expressed as a rise or fall in sea level, was something in the region of 1 cm per year. During the episode of global warming that we are experiencing today such changes are of obvious concern to people who are living at very low altitudes close to the present sea level.

For most of that part of geological history with a fossil record adequate for global correlation, namely the Phanerozoic, there is no evidence of polar ice caps. We cannot therefore plausibly make recourse to what is nowadays called *glacio-eustasy*. For much, if not most of, the twentieth century little attention was paid to any other kind of eustasy, partly because biostratigraphic correlation was neither sufficiently refined nor sufficiently extensive across the world, and partly because, in the absence of polar ice, there was no plausible mechanism. (There are perhaps echoes here of the continental drift controversy.) By the 1960s, however, the situation had improved in both these respects. In particular, a plausible mechanism emerged within the context of plate tectonics. The ridges that by then had been well mapped on the ocean floor were not just associated with the splitting and moving apart of continents. It was realized that their uplift was bound to displace sea water on to the continents, giving rise to marine transgressions that could be recognized in the stratigraphic record of those continents. To stick to the example mentioned earlier, the mid-Cretaceous transgression could unequivocally be shown to be global in extent. By plotting the distribution of marine deposits across the continents it could be estimated that at its maximum extent the transgression inundated approximately 40 per cent of these continents. The British Isles, for example, was completely flooded in the latter part of the Cretaceous period by the Chalk sea. The rise in sea level that then took place can be estimated by using the *hypsometric curve*, which shows in graphical form the proportion of the Earth's surface that is elevated in relation to a datum (sea level). After making a few reasonable adjustments, an estimated rise in sea level of about 200 m above the present position has been arrived at. Another approach is to assume that the rise in sea level is due entirely to the displacement of sea water on to the continents by the rise of volcanic ridges and the Pacific and Caribbean plateaux, whose Cretaceous age has been determined both by biostratigraphic and radiometric means. Calculating their changed volume then gives a best estimate of 250 m. These two estimates, determined by entirely different methods, give encouragement to the view that a rise of about 200–250 m can be taken as well established.

Volcanic ridges and plateaux can obviously subside as well as rise. Subsidence gives rise to regressions of the shallow seas that cover the continental shelves, the *epicontinental seas*, deposition from which has given rise to a very large proportion of the stratigraphic record. The corresponding rises and falls of sea level produced by this means are known as *tectonoeustatic* changes. The rate of change is of the order of 1 cm per 1000 years, about three orders of magnitude slower than glacioeustatic changes. There are nevertheless numerous cycles of change in the stratigraphic record which appear to indicate shallowing and deepening of the sea in a given area.

The origin of these relatively short-term changes is more controversial. Sedimentary cycles of this kind in the Carboniferous have commonly been attributed to glacioeustasy, because of the presence at that time of a polar ice cap in the southern hemisphere. The problem is that very similar cycles occur throughout the stratigraphic record (for example, in the Triassic and Jurassic), at times for which evidence of a polar ice cap is missing, while the distribution of the fossil fauna and flora suggests relative global equability. The biostratigraphic resolution available is rarely sufficiently refined to establish global correlation; and another handicap is that some of the cycles themselves, as opposed to the fossils, cannot be traced further than limited regions. Some geologists nevertheless insist on eustatic control, while others see the control as regional tectonism involving subsidence and uplift. For example, Mike Gurnis, of the California Institute of Technology, has explored the role of mantle dynamics in causing upward and downward movements of parts of the Earth's surface. He finds that Australia behaved counter to the world trend in the middle and late Cretaceous, by exhibiting uplift and regression. Gurnis is able to explain this by a rather elegant model involving the mantle.

The mechanism required to produce short-term changes, of the sort that could cause drastic environmental changes in a limited period of time, and hence perhaps be relevant to mass extinctions, continues to be controversial. Some geophysicists think that lateral stresses in the *lithosphere* could provide an explanation. The lithosphere is the rigid outermost layer of the Earth. It comprises both the crust and upper mantle, and its thickness varies from about 4 km under the oceans to perhaps as much as 300 km under parts of continents. Plate-tectonic forces could, it is thought, produce sharp vertical changes in the substratum on which sediments are deposited, but the great majority of these are likely to be regional rather than global in extent, rather like the tectonic movements investigated by Gurnis. Such regional movements could well cause local extinctions as habitats were drastically altered, but it is hard to see why the survivors could not find extensive areas of refuge, from which they could expand their ranges again once conditions had been restored later in the cycle.

Although the subject is a large and complex one, enough has been said to provide the background for a consideration of the possible relationship between sea-level changes and mass extinctions in the marine realm.

Newell's hypothesis

As mentioned in Chapter 3, Newell was the first person to suggest an explicit relationship between mass-extinction events and major falls of sea level giving rise to regressions within epicontinental seas. Today's continental shelf, at a time of relatively low sea level in relation to the whole of the Phanerozoic, could be called a pericontinental sea. In the geological past, shallow seas less than 200 m deep occupied a large proportion of the continental area. Seas of this type are usually called *epicontinental*. In contrast to the oceans proper, in an epicontinental sea a regression of quite modest dimensions can give rise to a substantial reduction in habitat area for the *benthos*, the marine organisms inhabiting the sea bed, and also to a lesser extent the *nekton*, those organisms swimming in the water above. Although representatives of all major phyla (categories of animals) occur in the deep ocean, sometimes in surprisingly high diversity, the great bulk of organisms, the *biomass*, live in shallow pericontinental seas and on the upper part of the continental slope. Shallow epicontinental seas, usually with low tidal ranges, must have been more vulnerable to environmental fluctuations with respect to such factors as temperature and salinity than the more 'buffered' (and therefore stable) ocean system, and a significant reduction in the area of their habitat could have been expected to have a deleterious effect on organisms that were well adapted only to a different environment. At the present day, reduction in habitat area, such as occurs when tropical rain forests are destroyed, is thought to be much the most important reason for the greatly increased extinction rates recognized in a wide variety of organisms.

For such reasons as these, Newell's hypothesis appears plausible. He proposed that his six marine mass extinctions were caused by reduction in habitat area as a result of the regression of epicontinental seas. The hypothesis needs, however, to be put to the test, and we have to ask whether the extinction events correlate with regressions produced by falls in sea level, as deduced from the stratigraphic record. As indicated above, the relevant information must come from three sources: from thorough biostratigraphic analyses across the world to establish the best possible correlations; from equally thorough facies analyses to determine shallowing-up and deepening-up successions; and from palaeogeographic analyses in which the changing distribution of land and sea are plotted through time.

Figure 5.3 displays my tentative sea-level curve for the whole Phanerozoic, with Newell's six mass-extinction events

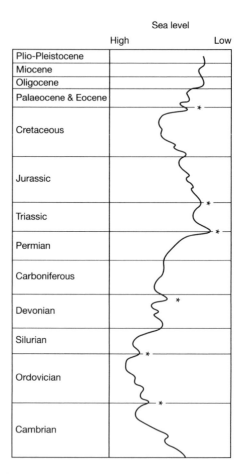

Fig. 5.3 The location of Newell's six major episodes of marine mass extinction, marked by asterisks, with respect to Hallam's first-order Phanerozoic sea-level curve. (After Hallam and Wignall (1997).)

asterisked. Sea-level curves are usually portrayed in this form, as squiggly lines, on a wide range of scales. Self-similarity probably applies, in that the amount of squiggliness does not change with scale, just as is true of coastlines. In other words, the system obeys fractal laws. We need to bear in mind, however, that finer scale sea-level changes, of shorter duration and lesser magnitude, are in many instances controversial, being no more than tentative and (as usual) qualitative models of rates and amounts of sea-level change, and that they reach a limit which depends on the refinement of the biostratigraphic control. Even the large-scale curve of Fig. 5.1 is best treated only as a broad approximation: it is based upon a review of existing stratigraphic literature, and could certainly be challenged in detail.

So while a perusal of Fig. 5.3 at first suggests a striking confirmation of Newell's hypothesis, the stratigraphic data associated with the extinction events have to be looked at more closely and critically, taking account of recent research. Before trying to do this, at least for the five events recognized by Raup and Sepkoski, we must consider the strongest criticism made of the regression hypothesis. By far the best-known marine regressions are the dramatically rapid ones of the Quaternary, which are associated with the locking-up of polar ice at the expense of sea water in times of cooler climate. Yet no associated extinctions appear to have taken place in the marine benthos, and this is apparently a clear contradiction of the hypothesis. There is, however, more than one answer to this criticism. The sea-level fluctuations of the Quaternary have been very brief in terms of the geological timescale, and it can be argued that organisms could readily re-expand from limited refugia when the sea level subsequently rose again.

This is because it is necessary to hold organisms in a stressed state for a minimum time before they become extinct, and the time may simply have been insufficient. Furthermore, Jablonski has argued from the evidence of possible island refugia in the oceans that Quaternary regressions would have caused only limited extinctions.

It can also be argued that the Quaternary was a very unusual time in Phanerozoic history, and those organisms that had survived until this time probably had a wide tolerance to environmental fluctuations. The older epicontinental seas, in which organisms lived that produced the vast bulk of the marine fossil record, might have been environmentally stable for long periods, such that even slight changes could have had a deleterious effect. We recall Ager's dictum, quoted in Chapter 2, about the stratigraphic record: 'Long periods of boredom interrupted by moments of terror'.

Modern interpretations of sea-level change in relation to the 'big five' extinction events

Since I wish to consider the end-Permian event in most detail I shall not deal with the extinction events in order of age.

1. End-Ordovician

The end-Ordovician extinction event coincides with one of the best-documented episodes in the stratigraphic record of a regression that was quickly followed by a transgression. A major fall in sea level began in the penultimate stage (the Rawtheyan), and in consequence the marine strata of the final,

Hirnantian, stage are of very limited extent. The succeeding transgression began in the late Hirnantian and continued into the basal stage of the Silurian, which is associated with widespread flooding of the continents associated with the spread of deep-water graptolite-bearing shales in many regions of the world. Globally synchronous sea-level changes in the Ordovician–Silurian boundary interval clearly imply a eustatic signature, and the presence of contemporaneous glacial deposits in the high palaeolatitudes of the southern continent Gondwana provide a likely cause. The basal Hirnantian regression is thus probably a response to ice-cap growth over Gondwana, and the basal Silurian transgression presumably records its rapid melting.

A detailed study of the fossil record indicates that there were not one but two phases of extinction, the first associated with the regression, the second with the subsequent transgression. The basal Hirnantian crisis preferentially eliminated the trilobites and low-latitude benthos; the late-Hirnantian event exterminated a distinctive brachiopod fauna and many deep-shelf taxa. The loss of large areas of marine habitat is widely regarded as the principal cause of the first phase of benthic extinctions, in accordance with Newell's hypothesis, but the second phase is clearly associated with sea-level rise and therefore requires a quite different explanation. We shall return to this topic in the next chapter.

2. End-Cretaceous

There has been a widespread recognition since well back in the nineteenth century that the latest Cretaceous was marked by a major regression, but the possible relevance of this to the contemporary mass extinctions has received surprisingly little attention in recent years. Consider the case, for instance, in England, where the eroded top of the Chalk is overlain with a marked hiatus by early, but not basal, Palaeocene strata. Only in limited areas in Europe such as Denmark is there a continuous marine succession extending from the Cretaceous to the Tertiary. Such sections as these, and the now-numerous sections known from drilling beneath the oceans, enable us to learn more about what happened across the Cretaceous–Tertiary boundary.

Gerta Keller and her German colleague Wolfgang Stinnesbeck provide the only global review of sea-level changes across the K–T boundary that takes full account of modern work on these sections. They and other workers agree that there was a major fall in sea level in the late Maastrichtian, the youngest Cretaceous stage, which was probably greater than anything since the mid-Cretaceous. This fall in sea level was followed by a rise that started just before the end of the period and continued into the Danian, the oldest stage of the Palaeocene, with a minor interlude of fall and rise within the earliest Danian. Keller and Stinnesbeck reckon that the major end-Cretaceous fall took place within about a hundred thousand years and amounted to 70 to 100 m, but the justification for these quantitative estimates is open to doubt. The speed of inferred change indeed suggests glacioeustatic control, but in the absence of evidence for a polar ice cap at this time we must remain sceptical of this. Keller and Stinnesbeck challenge the popular interpretation (popular at least among impact supporters) that the coarse clastic deposits around the Gulf of Mexico are tsunami deposits related to the Chicxulub impact event, pointing out that they are of variable ages and frequently pre-date the K–T boundary. Furthermore these clastic deposits do not represent a single-event deposit laid down over a few days, but multiple events over a longer period, and are more plausibly interpreted as deposits formed when sea level was at its lowest.

We have already considered some of the marine biotic changes across the K–T boundary. It is evident that not all the gradual decline in diversity recorded by specialists working on numerous fossil groups, as reported in a paper by Norman MacLeod and many other authors in 1997, can be dismissed as a consequence of statistical shortcomings in collection (the Signor–Lipps effect). The decline is in fact likely to reflect prolonged environmental changes, including changes in sea level. Gradual change appears to have been followed by a catastrophic event at the end of the period, no doubt bound up with impact. It is, however, difficult as yet to disentangle the environmental effects of climatic and sea-level change during the Maastrichtian, a time of significant environmental change.

Regression of the sea should, one would think, be irrelevant to the extinction of continental groups such as dinosaurs. Dave Archibald has, however, argued that the major changes in habitat produced in North America as a result of the latest Cretaceous regression were a significant cause of dinosaur extinctions in the times up to the end-Cretaceous catastrophic event that finished off the survivors, which he attributes to bolide impact.

3. End-Triassic

Some of the best evidence of an extensive fall in sea level, quickly followed by a rise, manifested in a variety of facies, is at the Triassic–Jurassic boundary. The evidence may, as in Bavaria, be in the form of channels at the base of the Jurassic with river deposits cut into marine Rhaetian, the topmost stage of the Triassic, or, as in Austria, be karstified marine limestone surfaces, which can be produced only by subaerial chemical weathering. In England it shows up locally as truncated limestone surfaces in the top Rhaetian and by pebbles of limestone in basal Jurassic shales eroded from Rhaetian limestones. This evidence, though less striking than what is seen elsewhere in Europe, is noteworthy because it occurs in an otherwise continuous marine succession. Throughout northern Europe there is a significant hiatus between the Triassic and Jurassic because the topmost Rhaetian has been removed by erosion. Everywhere in Europe there are clear indications from the facies displayed at the very base of the Jurassic or lower Hettangian stage of a pronounced marine transgression and deepening of the sea.

Elsewhere in the world the evidence is mixed. In various parts of North America, as in Arctic Canada and Nevada, the facies succession also signifies a shallowing–deepening couplet across the system boundary, but on the deeper-water western margins of the two American continents no such manifestation is evident. Marine sections across the T–J boundary in the other continents are very few, and no clear picture of any change in sea level emerges. Thus it cannot be held as conclusively established that the sea-level changes in question were genuinely eustatic, as opposed to being related to very widespread uplift and subsidence centred on the North Atlantic region, although eustasy remains a distinct possibility. The rate of change of sea level, whether regional or global, can be estimated on various grounds to have been far more rapid than would be associated with normal tectonic movements. There can be no question of invoking glacioeustasy, however, for there is no evidence of polar ice caps at this time, and every indication from other criteria of only a mild fall in temperature from the tropics to the poles.

The most striking organic change seen in the T–J boundary record is the almost total extinction of the ammonites, which had flourished and risen to a high level of diversity in the late Triassic after radiation of the survivors of the end-Permian debacle. Possibly no more than one genus survived into the Jurassic, from which radiated an even more impressive diversity of taxa in that period. It is a salutary thought for Jurassic biostratigraphers that their work would have been rendered considerably more difficult if the end-Triassic extinction of the group had been total, because the ammonites are immensely better than any other group of Jurassic fossils for correlation and provide the standard zonation everywhere. I have an especial partiality for the rocks and events of this period, and much of my palaeoenvironmental research involving fine correlation across the world would simply not have been possible without recourse to the information supplied by these splendid creatures.

The key question here, however, is how catastrophic was the extinction of the ammonites? Some general diagrams give the impression that their demise took place in a mere geological instant, but a close look at the record suggests otherwise, because many taxa disappeared well before the end of the Triassic. The Rhaetian faunas thus have a much lower diversity than those of the preceding two stages, which suggests a decline extending over several million years. A similar gradual decline is shown by those curious-looking microvertebrates the conodonts, which are well-nigh incomprehensible to all but the specialist. Having robustly survived the much greater extinction event at the end of the Permian, they persisted with steadily declining diversity through the Triassic, until their final demise, in Europe at least, came with the extinction of just one remaining genus.

The best record of marine macroinvertebrates, in terms of both diversity and individual abundance, comes from the bivalve molluscs (clams and oysters). Humbler they may be than dinosaurs, but they are far easier to find and much more amenable to statistical studies. They underwent significant extinctions during the T–J boundary event that seem to have been concentrated at the end of the Rhaetian, unlike the extinction of the ammonites and conodonts. It has to be added that the changes in facies up the stratal succession that are bound up with changes in sea level make it difficult in places to disentangle the disappearances caused by local environmental change from those that are effectively global in extent. Late Triassic reef ecosystems composed substantially of corals and calcareous sponges are wonderfully displayed in the Northern Calcareous Alps of Austria, especially on the Dachstein mountain, and disappear with dramatic suddenness at the very end of the period. The overlying basal Jurassic strata are, however, of a much deeper-water facies, because there was not only an important global rise in sea level at this time but also considerable tectonic subsidence in the Alpine region, probably bound up with the initiation of new ocean in the Mediterranean region. Nowhere else in the world are such rich Triassic reef faunas known, and the T–J boundary indeed marks an important extinction phase, especially among the sponges. We can be sure, though, that there was a catastrophic event only in the Alpine region, and this seems as much bound up with regional tectonics as anything.

4. Late Devonian

Modern studies have revealed that most of the latter part of the Devonian was characterized by elevated extinction rates with the intervals between the two youngest stages, the Frasnian and Famennian (F–F) and Devonian–Carboniferous (D–C) standing out as extinction peaks. It is the F–F event that is now regarded as one of the 'big five'. Newell originally postulated that the event may have been abrupt, and many subsequent researchers have concurred. It has, however, become evident that for many organic groups there was a protracted crisis spanning several conodont zones, with the peak of extinctions exactly at the F–F boundary. The victims of the F–F crisis included many shallow, warm-water benthic taxa, such as various brachiopods and trilobites, and most reef taxa (stromatoporoids, corals and foraminifera). Pelagic taxa living in the water column, such as conodonts, ammonoids and placoderm fish, were also severely affected. Deep-water and cold-water taxa appear to have survived the crisis better than others, with the result that they became common and widespread even in low-latitude Famennian sites. This 'aftermath fauna' includes hexactinellid sponges and high-latitude brachiopods.

Sea-level changes at the time of the F–F crisis have been widely perceived to have been rapid, of large magnitude, and in some way implicated in the extinction event. This, however, is the only broad consensus that exists in a subject of considerable debate. While the regression–extinction link continues to be popular in many F–F extinction scenarios, there is only equivocal sedimentological evidence for what would in geological terms have been high-frequency oscillations in sea level during the F–F crisis. The extinctions peaked precisely at the

F–F boundary at a time when the relative sea level in some regions was either at its highest, and therefore changing little, or was during a transgressive phase and rising rapidly. Other environmental parameters such as cooling and anoxia are probably more salient factors to consider in F–F extinction mechanisms.

Other, lesser extinction events associated with regression

Until recently the later part of the Permian has been regarded as an interval of protracted crisis. However, as reviewed by Hallam and Wignall in their 1997 book, new research has revealed that the interval encompasses two distinct extinction events, separated by an interval of radiation and recovery. The first event occurred near the end of the Guadaloupian stage, early in the late Permian and several million years before the end of the period. Low-latitude benthic faunas were particularly hard hit, and many fusulinid foraminifera, echinoderms, brachiopods, and bryozoans were among the victims. The end-Guadaloupian has long been recognized as a major interval of regression. In west Texas, for example, the regression terminated the growth of the celebrated Capitan reefs and marked the end of normal marine deposition in the region for the remainder of the Permian. In South China there was an equally severe base-level fall at this time, which suggests that the event was probably of global rather than merely regional extent. Elsewhere in the world the record is too poor, usually because the sedimentary deposits of this age are continental, to confirm or deny this proposal. Most studies of this crisis invoke the loss of marine habitats as a significant cause of extinction.

Another notable regressive event, the Hawke Bay Event, took place late in the early Cambrian. It is thought by the Russian palaeontologist A.Y. Zhuravlev to have been worldwide in extent. It was, according to Zhuravlev, sufficiently severe to have caused the extinction of more than 50 per cent of the benthic genera extant at this time, and led to the severe elimination of reef biota, mainly the archaeocyathid sponges.

5. End-Permian

The most important extinction event of all has been left to the last because I have had some personal involvement in research on variations in sea level across the Permian–Triassic boundary, and I wish to indulge myself a little in giving the reader some idea of what is involved in doing fieldwork in remote parts of the world that are never visited by the tourist. While geology, like other sciences, has benefited enormously from technological advances in learning about the structure and distribution of rocks by using remote sensing methods, much can still be learned, and needs to be learned, using time-honoured methods. The geologist's traditional tool of the trade, the hammer, remains as important as ever to those of us who are not always working in laboratories or peering at computer screens. In the words of Samuel Butler, 'There's many a good tune played on an old fiddle.'

The traditional view accepted by most geologists has been that the end of the Permian period marked the regressive peak of a temporally extended major fall in sea level that lasted for some time before a shift to rising sea level in the earliest

Triassic (see Fig. 5.3). It was accordingly natural for many geologists to agree with Newell that significant reduction in habitat area was at least the principal cause of marine benthic extinctions at the end of the Permian. Consider the evidence in the south-western United States, which has the best marine Permian record in the subcontinent. The marine Permian there culminated with the growth of the celebrated Capitan reefs, the youngest, post-Guadaloupian strata being entirely non-marine. This pattern is characteristic of most of the world, but in recent years our understanding has changed dramatically as a result of work in south China, which turns out to have a large area revealing a continuous marine transition from the Permian to the Triassic. The research used there has entailed biostratigraphic zonation, using conodonts, and intensive examination of strata.

As mentioned above, there was indeed a major regression at the end of the Guadaloupian, which seems to have been the prime cause of many extinctions at that time, but in many parts of the world the sea had advanced once more over a large part of the continental area millions of years before the end of the Permian period. What actually happened at the P–T boundary therefore required detailed facies analysis, using the new conodont biostratigraphy, to explore the possible environmental changes across that boundary.

Paul Wignall became a research student of mine in the mid-1980s and quickly acquired authoritative knowledge on the palaeoecology of oxygen-deficient benthic environments. After his graduation I naturally wished to keep in touch with a person of his exceptional ability. Since we got on well and had closely similar interests, we applied successfully for grant money to examine in detail strata in three parts of the world where there were excellent marine sections across the P–T boundary: the Alps, the Salt Range of Pakistan, and southern China. The intention was not only to draw up our own sea-level profiles but to investigate evidence of possible oxygen deficiency, a subject that I shall discuss in the next chapter.

The Alpine sections occur in the Dolomites, an area with the most glorious scenery in the whole of Europe, if not in the world. Non-geological friends are often rather cynical about the places we choose for fieldwork, but there can be pain as well as gain. Mountains have the great advantage of abundant rock outcrop, but to reach the best places often entails labori-ous and occasionally dangerous climbing. Some of the more difficult spots I had to leave to Paul, who is as agile as a moun-tain goat, but for the most part the sections were readily acces-sible without exerting too much effort, and one classic section was conveniently by a roadside. The local stratigraphy, with-out going too far into detail, is that the top of the Permian has been taken as the summit of the Bellerophon Formation, capped by a limestone, and the base of the Triassic at the boundary between this and the overlying Tesero Oolite. This boundary is marked in places by a clear erosional horizon: the sort of evidence that is universally accepted as indicative of regression. Stratigraphers today would usually refer to it as a *sequence boundary*. The conodont evidence shows, however, that the P–T boundary lies at the top, not the base, of the trans-gressive Tesero Oolite. Other Permian fossils also occur in the Tesero Oolite, and do not disappear from this limestone until the start of the overlying shale, which belongs to the Mazzin Member of the Werfen Formation. There is thus no correlation in the Dolomites between regression and extinction, and the Newell hypothesis is not supported.

A similar situation is recognizable in the Pakistan Salt Range, which lies next to the Indus Valley south of Islamabad, the capital. The Salt Range Permian has long been a classic system for stratigraphers. It has yielded rich faunas of brachiopods and other important benthic groups that disappeared at the top of the system so dramatically that Schindewolf was led to invoke a burst of cosmic radiation to produce the mass extinction. The Chhidru Formation is the youngest Permian unit and is overlain erosively by the Kathwai Dolomite. The P–T boundary has traditionally been taken at the boundary between these two stratigraphic units. Once again, in this shallow marine succession, Permian fossils ignore the sequence boundary and continue into the Kathwai, disappearing only in the overlying shales.

Our visit to the Salt Range would not have been possible but for the generous help offered by the Pakistan Geological Survey, whose deputy director, Mahmoud Raza, accompanied us in a Toyota Land Cruiser belonging to the Survey. One day in the field, Paul and I were measuring a section in a gully when we looked up the hillside to see Mahmoud engaged in conversation with a tall, slim young man dressed in the traditional *shalwa kameez*, with a rifle over his shoulder – not an unusual sight in the remoter uplands of Pakistan. Within a few minutes a disturbed-looking Mahmoud joined us and said that we were getting out of here, in a tone that brooked no dissent. As we sped along the road leading to the small town where our hotel was situated, he gave the reason for his precipitate decision. The man he had encountered on the hillside had introduced himself as someone who was on the run from the law and so was obliged to live in the mountains away from any serious attempt by the police to capture him. He then begged for some money. When Mahmoud politely demurred, the man began to finger the trigger of his rifle as he repeated his request, after which Mahmoud, deciding that discretion was the better part of valour, promptly emptied his wallet of rupees and departed promptly to rejoin us.

Fortunately the next day the District Commissioner was holding court to a gathering of his senior officials on one of his regular visits to the town. We were granted audience at the end of the meeting, with everyone still present. When we explained what had happened up in the Salt Range, the D.C. turned to his Chief of Police and said, 'Give them a couple of your men for the next week.' After we took our leave, expressing our gratitude but stopping short of bowing, the D.C.'s parting shot was, 'What do you want to go up into the Salt Range for? That's country fit only for savages.' One was tempted to reply, but didn't, 'Well, savages and geologists'.

So the next day we set off to renew our fieldwork in the company of two burly policemen equipped, naturally, with rifles. Needless to say, no more sightings were made of the savage, or dacoit, and all the policemen had to do was hang around and chat with each other and the driver, or very occasionally with a passing shepherd. Still, they seemed quite pleased with their unusual assignment, which was no doubt a welcome change from their normal routine.

The Pakistan Geological Survey's headquarters in Islamabad had provided me with a fascinating insight into state bureaucracy in the developing world during an earlier visit to the Salt Range with the American geochemist Barry Maynard. To acquire relevant permissions for fieldwork we had to run the gauntlet of successive visits to progressively more senior officials. At last we reached the most senior official, clean-shaven, smartly dressed in a western-style suit, and occupying an appropriately large office. At his left sat a bearded man in traditional dress, who sat silently while the rest of us conversed (almost certainly he knew no English). His presence in the office was indeed a mystery, until the telephone rang. He immediately picked it up and handed it to his boss, thereby identifying himself as the telephone wallah. No doubt he could go back to his wife in the evening, telling her that he had had a busy day at the office. This was to us a classic example of overmanning by the state. In recounting this anecdote I have absolutely no wish to be patronizing. Given an exceptionally high unemployment rate, the State deems it desirable to give as many people as possible some kind of employment, no doubt at very modest rates of pay.

Let me continue my digression a little further by saying something about the drivers of the field vehicles used in the developing world, which are usually Toyota Land Cruisers rather than Land Rovers which, though they usually last longer, are considerably more expensive. The Pakistan Geological Survey drivers are marked out sartorially from their 'social superiors'. (I'm sorry, but that's the truth.) Whereas the scientists and administrators wear normal western clothes, the drivers and other 'humbler' staff are dressed in *shalwa kameez*. My use of the adjective 'humbler' in the case of the drivers is meant to be ironic, because they are often very skilled and resourceful people on whom the rest of us on field trips are very dependent. This was brought home to me some years ago in Chile, not a country considered part of the developing world, and more resembling one of the less wealthy southern European countries. I was with the Chilean Geological Survey engaged in fieldwork in the Andes, which entailed camping in a remote area. Our driver (in a Land Cruiser, of course) was a *mestizo*, unlike the geologists, all of whom were solely of European stock. On our first day out from Santiago, he negotiated with admirable skill and fortitude some very tricky mountain tracks. Eventually the time came to set up camp. I did the normal English thing and offered to help, but my Chilean geologist colleagues beckoned me back. The driver then continued on his own to erect two tents with breathtaking efficiency; but that was not all. He then proceeded to cook a very acceptable supper for us all. The geologists explained that this was all part of his job, the pay for which was, needless to say, much lower than theirs. Though the driver seemed jolly enough, I was left feeling rather uncomfortable about the existence of what seemed somewhat Victorian values in an otherwise delightful country. For those who may still berate the English for their obsession with class I am inclined to proffer the Victorian riposte 'stuff and nonsense'. Our class system, though it still exists, is vastly weaker than it was even two generations ago, and it pales in comparison with that which is still exercised in a large part of the world.

As for resourcefulness, I shall recount a further anecdote involving a recent field trip undertaken by Paul Wignall and myself to southern Tibet in the shadows of the Himalayas. We were extinction-boundary hunting as usual. Our Tibetan driver was unusually capable in negotiating some very difficult tracks in the mountains. In places he had to fill in chasms in the trackway with stones before we could proceed further. One stretch was more terrifying than either of us had ever experienced before, and Paul was poised to leap out of the Land Cruiser to avoid tumbling into a gorge. It was on our way back to Lhasa, however, that another type of resourcefulness was called upon. On reaching a small town, we needed to refuel because the tank was almost empty. There was only one petrol station in town. To our dismay we were told that it had run out of fuel, and the next delivery from a tanker was not due for several days. But we had to reach Lhasa the next day in order to fly back to England via Chengdu and Hong Kong. With the complications and extra expense that could ensue from any delay we faced a possible calamity. So what did the driver do while we were looking helpless and frustrated? He negotiated with a local restaurant owner for some petrol, having to pay about twice over the odds. With that fuel we were able to continue to a larger town with several petrol stations and our problems were over. Now who would have thought of approaching a restaurant for petrol? It just happens to be the standard wok-heating fuel in some Tibetan restaurants. Paul and I poked our nose into one such kitchen, and an alarming sight it was.

It is time to return to geology. After our visits to the Alps and Pakistan, Paul and I wished to examine P–T boundary sections in a region of deeper-water basinal facies, namely the classic one at Meishan, near Nanjing in southern China, and an excellently exposed one in a much less accessible location, close to the village of Shangsi in the mountainous north-western part of Sichuan Province. The Meishan section has been studied in great detail by Chinese palaeontologists, who recognized a dramatic disappearance of species exactly at the boundary as determined by conodont biostratigraphy. It also coincides with a sharp fall in the carbon isotope ratio that resembles the one at the K–T boundary. This is a phenomenon that seems to characterize the P–T boundary globally. The standard interpretation of the carbon isotope shift has been in terms of a fall in productivity produced by mass extinction, giving rise to a 'Strangelove Ocean'. We shall see, however, in Chapter 7 that an alternative interpretation is possible. The Meishan section contains numerous volcanic ash layers inter-bedded with the shales and limestones, and radiometric ages can be obtained from these ash layers. Using the radiometric ages it can be inferred that the mass extinction took place within half a million years at most, and probably much less.

Our facies investigations gave no evidence of water shal-lowing, implying a fall in sea level, culminating at the P–T boundary, as might be expected from Newell's regression scenario. Instead, there is evidence of deepening, correspond-ing to a marine transgression, beginning in the latest Permian and continuing into the Triassic. The picture is identical in the section near Shangsi, a village so remote that the locals stared at the two of us as though we were visitors from another planet. In the absence of television they had evidently never seen Caucasians before, let alone a blond one, as in Paul's case. (All the Han Chinese, without exception, are raven-haired.)

The regions we had visited represent the best places to study a continuous marine section across the P–T boundary in what was in Permo-Triassic times the low-latitude zone, but there are also comparably good high-latitude sections in Spits-bergen and East Greenland. Paul has subsequently visited these with a number of colleagues, and their collaborative research clearly indicates a sea-level change picture com-parable to that seen in low latitudes. The event is therefore clearly global in extent, and the pronounced eustatic rise is one of the most striking to be recognized in the whole stratigraphic record, although unfortunately it is impossible as yet to pro-duce any reliable quantitative estimate. The East Greenland

P–T boundary section is the most stratigraphically extended of all. By comparing the fine biostratigraphic zonation with Meishan, which has been precisely dated radiometrically, an estimate of about 10 to 60 thousand years has been made for the duration of the mass-extinction event. This is certainly a catastrophe in geological terms. Furthermore it can be shown, from the collaborative work on conodonts and on pollen and spores carried out to sea by the wind, that marine and continental extinctions correlate precisely.

Conclusion

It should be evident that, while there is a strong association between marine mass extinctions and sea-level changes, it is generally less with fall in sea level and the associated regression, as proposed by Newell, than with rise in sea level. But why should marine organisms be adversely affected by a rise in sea level, which would lead to an expansion rather than a decrease of the area of their habitat? That another factor is involved, which correlates with sea-level rise, will be explored in the next chapter.

6

Oxygen deficiency in the oceans

We are all very much aware that oxygen deprivation leads quickly to death, and this is true not just of our own species but of virtually the whole organic world. There are indeed very few exceptions, such as the anaerobic bacteria that derive their energy from reducing sulphates to sulphides, which flourish in the absence of free oxygen. (As these organisms do not leave a fossil record they provide no clues for the geological detective.) Today the atmosphere never lacks oxygen, except in artificially enclosed conditions, but oxygen deficiency can be lethal in certain marine environments and thus must be explored as a possible factor in causing mass extinctions.

Mixing with atmospheric winds ensures that the surface waters of the ocean, down to the greatest depth attained by storm waves, always contain plenty of oxygen. Most of the oceans and marginal seas today contain oxygen throughout their depth, but in certain circumstances an oxygen deficiency can occur in the lower parts of the ocean. In parts of some tropical oceans, for instance, the oxygen content decreases with depth until near the ocean bottom, where under the influence of currents driven by cold water from around Antarctica, the oxygen content increases again. This gives rise to a zone in the ocean known as the *oxygen minimum zone*. The rapid deep ocean circulation is today driven ultimately by the presence of polar ice on Antarctica, which is the main cause of the strong sea-water temperature gradient from the tropics to the poles. For long periods in the Earth's history substantial polar ice caps were lacking, and many geologists believe that during those periods latitudinal ocean currents were more sluggish. The deep ocean must then have been largely deficient in oxygen, if not completely lacking in oxygen (*anoxic*). (Sea water with a content of one or more millilitres of oxygen per litre of water is called *oxic*; 0.1 ml or less is *anoxic*; and for any intermediate value the water is *dysoxic*.)

Certain parts of the sea bed where the overlying water is deficient in oxygen are enriched in organic matter derived principally from the plankton. This is because not enough oxygen is present to facilitate normal aerobic decay. There are two schools of thought about the cause of this organic enrichment. On the one hand, the *high-productivity model* favours high primary productivity stimulated by nutrients as the origin of the organic enrichment. In this model, low-oxygen (dysoxic) conditions in the bottom water may occur because of a higher oxygen demand caused by the decay of the organic matter; the dysoxia is, however, viewed as a consequence, not a cause, of the increased amount of organic matter. In contrast, the *preservation model* envisages lack of oxygen (dysoxia or anoxia) as the cause of enhanced preservation of organic matter under conditions that are not necessarily highly productive. The rates at which dissolved oxygen is supplied to bottom water are usually considered to be inhibited by the presence of a strong density interface (a *pycnocline*) within the water column. This interface can be the consequence of stable stratification of the water column, with minimal mixing. Stable stratification can be the result of differences in salinity or of differences in temperature; waters of lower salinity and waters at higher temperatures are both less dense than saltier or cooler waters. In the Black Sea, for example, the influx of fresh water from major Russian and Ukrainian rivers results in the development of a change in salinity with depth (a *halocline*). A stable stratification is ensured by the fact that the shallow sill of the Bosphorus inhibits free circulation with the Mediterranean. There are many other marine settings, however, in which a zone of rapidly changing density and temperature, a *thermocline*, has developed in stably stratified waters because of contrasts in temperature. Where a substantial part of the water column is anoxic it is termed *euxinic* (a word derived from the ancient Greek name for the Black Sea).

Nearshore anoxia is often developed during the summer months in the deeper waters of modern shelf seas, when water trapped beneath a summer pycnocline (either a thermocline or halocline) becomes deoxygenated as a consequence of the decay of plankton blooms. The effect can be particularly intense when productivity is stimulated by the introduction of high levels of nutrients of human origin, or *anthropogenic nutrient influx*. A significant consequence of such an influx may be the mass killing of organisms living on the sea bottom, the *benthos*, or those swimming close to the sea bottom, the *nektobenthos*. In normal circumstances oxygen is renewed in the winter months when storms cause strong mixing of the sea water. Life can then flourish again at the start of the next season. It is not difficult to see that in slightly changed circumstances such an annual renewal might not take place. If anoxia persisted for long enough over a sufficiently wide area, substantial extinctions could ensue.

How do we recognize evidence of marine oxygen deficiency (anoxia or dysoxia) in the stratigraphic record? The most characteristic deposits laid down in anoxic conditions are known broadly as black shales, although they need not necessarily be either black or shaly. Those with little organic matter may be no darker than grey, and some deposits are calcareous enough to be termed marls or even limestones. The most highly distinctive feature of black shales is fine sedimentary lamination, which may in some instances represent annual layers. Lamination of this kind is destroyed if oxygen is present in the bottom waters, because the oxygen supports benthic life, including a variety of animals that burrow in the sediment churning it up in the process. Whatever organic matter may originally have been present in the sediment is then for the most part either eaten or otherwise lost by oxidation.

Black shales are of especial interest to industrial geologists because they are the source rocks for the hydrocarbons that constitute liquid petroleum and gaseous methane, on which our society depends so heavily for its energy supplies. The solid compounds that comprise the organic matter in sediments are known collectively as *kerogen*. A lower limit of about 0.4 per cent organic carbon is generally considered necessary for a rock to act as a source bed, but most recognized source beds contain 0.8–2 per cent and the best as much as 10 per cent. The most important kerogen results from the accumulation and decomposition of large amounts of planktonic organisms. Where dead organic matter reaches the sea floor the constituents necessary for hydrocarbon generation are preserved only if the water is essentially anoxic. The change from organic matter to kerogen takes place after burial from shallow depths down to about 1000 m with temperatures of up to about 50 °C. With deeper burial and heating (at 1000–6000 m and 50–175 °C) the large molecules in kerogen break down to form smaller hydrocarbons of lower molecular weight. Oxygen is lost rapidly by dehydration and loss of carbon dioxide (CO_2) with the result that H_2O and CO_2 are the initial products. At higher temperatures volatiles and liquid petroleum products develop. In the appropriate structural setting the gas and petroleum will migrate through and across strata to accumulate in porous rocks known as *reservoir rocks*, to await discovery and extraction by the drilling of boreholes, either on land or on the sea bed.

Black shales have a great fascination also for palaeontologists. While they may be barren of benthos, active swimmers in the water column above the anoxic zone such as cephalopods, fish, and reptiles fall to the sea bed after death and are often exquisitely preserved in the sediment because there have been no scavengers on the sea bed to consume the flesh or disturb the bones. Nor have there been strong bottom currents to disturb the remains after burial. Thus most of the best examples of marine vertebrate fossils that can be seen in our museums come from black shales.

An excellent example is provided by the Lower Jurassic Posidonia Shales (Posidonienschiefer in German) of the German state of Baden-Württemberg. Superb specimens of the famous reptiles known as ichthyosaurs and plesiosaurs, together with marine crocodiles and many ammonites and belemnites, can be seen in Stuttgart's spectacular museum,

Fig. 6.1 An excellently preserved ichthyosaur, a swimming reptile, from the Lower Jurassic Posidonia Shales of Holzmaden, south-west Germany. Not only have the bones not been disarticulated by scavengers after death, but the outline of the original soft parts is preserved as a black carbonaceous film.

which is devoted entirely to palaeontology, as well as in the museum in the small town of Holzmaden, close to the famous quarries. Many of the ichthyosaurs have their full body outlines preserved as carbonaceous surfaces (**Fig. 6.1**). We can thus deduce the presence of fins and a tail outline not revealed by the skeleton alone. This special preservation testifies to a complete lack of disturbance between the settling of the cadaver and its ultimate burial. The vertebrates of the Posidonia Shales are no less famous than the beautiful fish from the Eocene Green River Formation of Wyoming, which appear in reputable fossil shops throughout the world. In this instance the deposit was laid down in a lake rather than in the sea. It is a light grey marly limestone rather than a black shale, but it also was deposited in anoxic conditions and demonstrates what have been interpreted as annual laminae.

The sedimentary record contains numerous examples of black shales. They are outnumbered by other types of deposit laid down in fully oxygenated bottom waters, with which they are frequently interstratified as a result of changing environments with time on or close to the sea bed. A number of sedimentological and palaeoecological criteria allow the distinction of oxic, dysoxic, and anoxic environments (**Fig. 6.2**). Well-oxygenated, or oxic, bottom waters are normally characterized by a diverse and abundant benthic life, including organisms living both on and within the topmost layers of the sea bed. Those living within the sediment may burrow deeply and produce thereby a variety of *trace fossils*. Disturbance of the sediment – *bioturbation* – is intense and any original sedimentary lamination is quickly destroyed. Less well-oxygenated, dysoxic, bottom waters have lower organic diversity and abundance; the depth within the sediment to which animals burrow is less; and the variety of 'infaunal' trace fossils is reduced. Progressively more poorly oxygenated conditions are signified by the disappearance of benthos and the appearance of lamination; the only fossils are those that have fallen in dead from the nekton or plankton.

There are also a number of mineralogical and geochemical criteria for determining the existence of anoxia in the water column at the time of deposition. Pyrite (iron disulphide, FeS_2) is the characteristic mineral of black shales. It has been formed by the interaction of iron with hydrogen sulphide, which is the product of sulphate-reducing bacteria operating in anaerobic (or anoxic) conditions. An important qualification must be applied, however, to the use of pyrite as an indication of anoxic sea water as distinct from an anoxic environment within the sediment. All fine-grained sediments of the sort that eventually lithify to shales or mudstones become anoxic at a short distance below the sediment–water interface because oxygen consumed by organic decay is not replenished and

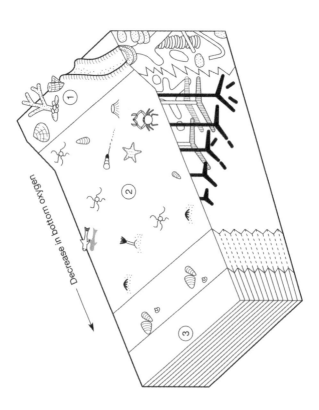

Fig. 6.2 Schematic representation of three oxygen-related biofacies. In order of decreasing levels of bottom-water oxygenation these are (1) oxic, (2) dysoxic, and (3) anoxic. Note the reduction of diversity, burrow diameter, and vertical extent of biogenic structures with decreasing oxygenation within zone 2. Zone 3 is characterized by laminated sediments poor in benthos; zone 2 by partly laminated or fissile sediments with a limited benthos: zone 1 by thoroughly bioturbated sediment rich in benthos, including a variety of active burrowers.

hydrogen sulphide is produced by sulphate-reducing bacteria. This can easily be recognized on tidal mudflats such as those in the Wash of eastern England. Below a thin brown surface layer containing iron oxide is a thicker zone coloured blue-black, from which emanates the highly distinctive and unpleasant 'rotten eggs' smell of H_2S. An important consequence of this is that pyrite can form in anoxic conditions some time after the deposition of sediment in normal oxic waters.

The key factor in distinguishing anoxic from oxic bottom waters is the size of the aggregates of pyrite crystals. In anoxic waters, there is a severe restriction on the size of the minute aggregates of what are known as pyrite framboids, which have formed within the water column and are thus a good indicator of euxinic conditions. Pyrite aggregates within fine-grained sediment deposited in oxic waters grow within the sediment to a much greater size and are often visible to the naked eye.

Iron geochemistry is also strongly controlled by what are known as *redox* conditions (a measure of the degree of oxidation and reduction) during deposition. Without going into unnecessary detail, an index known as the *degree of pyritization* can be used to determine the amount of oxygen-restricted deposition. The index cannot, however, generally distinguish between dysaerobic strata deposited under dysoxic conditions and euxinic strata formed beneath a sulphidic lower water column. Another inorganic geochemical measure is the amount of uranium in the sediment. Uranium and thorium emit gamma rays, and portable gamma-ray spectrometers can therefore be used in the field to measure the thorium and uranium concentrations in shales. A Th/U ratio of less than 2 generally signifies deposition in dysoxic or anoxic conditions.

This is because uranium is insoluble in highly reducing (anoxic) conditions but soluble in oxidizing conditions, whereas thorium is unaffected by redox conditions and remains insoluble. Organic chemistry also has its uses here. The chemical *isorenieratene* forms only in anoxic conditions within the photic zone, the region of the sea that is penetrated by light. Isorenieratene must therefore be formed only in shallow water.

Anoxic episodes and mass extinctions

While many black shales are recognized in the stratigraphic record, only a minority are correlated with mass extinctions. These are all extremely widespread and are usually associated with pronounced eustatic rises of sea level, which had the effect of causing the deeper, more poorly oxygenated waters to extend into the ocean margins. A localized anoxic event affecting only a limited region, such as a sedimentary basin, might well have had a severely deleterious effect on benthic and nektobenthic life, but it would not be expected to cause extinction because the region would be readily repopulated once conditions had ameliorated. The more widespread the anoxic event, the smaller would be the areas of refuge which could facilitate repopulation later. In a similar way, the more sustained the period of anoxia, the more likely a significant extinction would ensue.

We can now consider the 'big five' extinction events in terms of anoxic episodes and then go on to look at several lesser mass-extinction events.

Permian–Triassic boundary

When Paul Wignall and I began to explore the Permian–Triassic (P–T) boundary in Italy, Pakistan, and China one of our major objectives was to test our hunch that anoxia was at least heavily implicated in the marine extinctions that took place at that time. The evidence everywhere was so strong as to convert the hunch into a firm proposal. The tell-tale fine sedimentary lamination in the oldest Triassic strata was ubiquitous, as were the abundant and finely dispersed pyrite framboids indicating not only anoxic conditions in the bottom water but also euxinic conditions extending well up into the water column. The underlying Permian strata did not exhibit evidence of sustained anoxia, though black shale horizons occur intermittently in the deeper water basinal sections in China. What was missing in the basal Triassic were true black shales rich in organic matter; most of the shales were grey. A deficiency in organic matter in the bottom sediment would be a natural consequence of mass extinction of the plankton, and this is suggested by the pronounced negative shift of the carbon isotope curve at this horizon, as at the Cretaceous–Tertiary boundary. Paul's subsequent research in Spitzbergen and Greenland confirmed what we had encountered in low palaeolatitudes and provided final assurance that the phenomenon was truly global.

As would have been expected, the benthic fauna was virtually absent, except from a limited number of horizons, which were rarely more than a few centimetres thick and usually much less, crowded with the shells of a bivalve genus called *Claraia*. This genus is a characteristic representative of a group known informally as paper pectens, with shell valves that are only moderately convex. In life they lie on the sea bed. These fossils typically occur in associations of very low diversity – usually containing only one species – but of high individual abundance, so that the bedding planes are crowded. This unusual type of association indicates what is commonly called a high-stress environment, something that cannot be tolerated by most species. We interpret the high stress in this instance to be caused by oxygen deficiency. The shell beds mark relatively brief episodes when the environment ameliorated from anoxic to dysoxic, before reverting to anoxic. Other benthos is very rare, even in the shell beds, and most of the basal Triassic strata are barren of benthos. Some nekton, such as ammonites and fish, do occur however, showing that the entire water column was not anoxic. On the other hand the P–T boundary marks a mass extinction event for the planktonic radiolarian micro-organisms, raising the question of whether anoxia or some other factor was the cause of their substantial demise at this time.

All the regions that Paul Wignall and I have studied include strata that were deposited in relatively shallow epicontinental seas. What about the deep ocean? Unfortunately our knowl-edge of the deep ocean in Permian and Triassic times is extremely limited, because the ocean floor of this time has since mostly been subducted by plate-tectonic processes. However, in limited areas around the Pacific margin, traces of ocean-floor deposits have escaped subduction and have been preserved by accretion on to adjacent land. This has been clearly established, for instance, in Japan, where a series of what are known as radiolarian cherts, of Permian and Triassic age, occur in an accretionary prism – a wedge of sediments on the landward side of an ocean trench. Chert is a hard siliceous rock, equivalent to the flint of the English Chalk. Whereas the flint derives its silica from siliceous sponges, the Japanese rock has the protistan radiolaria as its source of silica. At the present-day, radiolaria dominate the deep ocean. Radiolarian ooze was first discovered during the *Challenger* expedition in the late nineteenth century. It is therefore very likely that the Japanese radiolarian chert represents a similar deep ocean deposit. The significant point to be made here is that the chert succession is interrupted at the P–T boundary by a black shale. There can be no better confirmation that the deep ocean, as well as the epicontinental seas, was anoxic at this time. It is therefore clear that the deep ocean could not have acted as a refuge. The situation today is different: the 'oxic' deep ocean contains many organisms that lived in shallow seas back in the Mesozoic, before they were displaced into deeper water towards the end of the era by newly evolved organisms, including teleost (bony) fish, carnivorous gastropods, and crabs. Some refugia must nevertheless have existed some-where, or else all the benthos would probably have become extinct; and this evidently did not happen. Where this 'some-where' is located is not readily revealed from the available rock outcrops.

Ordovician–Silurian boundary

In the previous chapter I indicated that the end-Ordovician extinction event was a twofold process. The earlier phase of the extinction correlates with widespread regression, which suggests a lowering of sea level, but the later phase is quite different, with strong evidence of a rise in sea level associated with an extensive spread of black shales in the earliest Silurian. These black shales are typically rich in graptolites, colonial planktonic organisms that were common in the early Palaeozoic. When they died, their remains fell to the sea bed.

Benthic life was substantially absent, however, because of oxygen deprivation. Major extinctions of graptolites and other planktonic groups, such as the acritarchs and the chitinozoans, took place during the earlier event and the earliest Silurian marks the beginning of radiation. Thus the monograptids began their spectacular Silurian radiation at this time. It is clear that the second extinction phase affected only the bottom waters; a spread of anoxia is the likeliest cause.

Frasnian–Famennian boundary (Late Devonian)

One of the world's classic late Devonian successions occurs in western Germany. The Frasnian–Famennian boundary coincides with what has been termed from this region the Kellwasser Horizons, consisting of thin beds of dark, organic-rich, laminated shales or limestones interbedded with shallow marine limestones. These horizons contain a distinctive fauna of molluscs and fish typical of Devonian euxinic basinal settings. Their occurrence in shelf settings records the brief expansion of such conditions into shallow-water sites. Evidence of a rise in sea level and the associated spread of anoxic or dysoxic facies is not confined to Germany, and is strongly suggestive of a global or eustatic event. In north-west Australia the classic reef facies (corals and stromatoporoids) of the Canning Basin is abruptly succeeded by deeper-water thin-bedded limestones containing ammonoids, but it is disputed whether or not these signify anoxic conditions. Anoxia has long been one of the most plausible candidates for the Kellwasser extinctions. The near-conclusive evidence includes:

(1) the widespread development of anoxic and dysoxic facies, corresponding closely with the extinctions;

(2) the preferential survival of dysoxia-tolerant benthos, such as certain ostracods and bivalves, and many deeper-water taxa among the sponges, rugose corals, and ammonoids, which were probably able to withstand low oxygen levels;

(3) diverse geochemical evidence, such as sulphur isotope values indicating the burial of large amounts of pyrite, increases in carbon isotope concentrations, indicating the burial of large amounts of organic carbon, and the enrichment of certain trace metals in the boundary sediments.

Triassic–Jurassic boundary

Unlike the earlier examples, there is no evidence of an extensive anoxic episode at the beginning of the Jurassic, although there was a well-authenticated rise in sea level, which was probably eustatic. In northern Europe, which has some of the best stratigraphic sections, there are positive indications of anoxic and dysoxic facies in the oldest Jurassic strata (Lower Hettangian stage), but the extinctions had apparently taken place through the youngest Triassic in response to the preceding regression.

Cretaceous–Tertiary boundary

As for the Tertiary–Jurassic boundary, bottom water anoxia does not appear to have played a notable role in the K–T boundary extinctions, although locally, as in the global type section in Tunisia, and in sections in both Europe and New Zealand, the basal Tertiary is marked by a thin black shale that is correlated with a rise of sea level.

Let us turn now to a number of events which, although less important than the 'big five', are sufficiently notable to warrant the term 'mass extinction'.

Early Toarcian (Early Jurassic)

The stratigraphic equivalent of the Posidonienschiefer in England is another black shale, excellently exposed on the Yorkshire coast near Whitby, where it is known as the Jet Rock. This takes its name from a black shiny material derived from the lignification of driftwood. The ease with which this material can be carved made possible the creation in Victorian times of a small industry in manufacturing ornaments such as necklaces, which can still be obtained from shops in the old part of Whitby.

The Jet Rock is a beautiful example of a finely laminated black shale, and has received intensive palaeontological, sedimentological and geochemical study. The sparse benthic fossils include at the base a thin layer rich in the 'paper pecten' *Bositra,* whose old name, *Posidonia,* gives its name to the Posidonienschiefer of Germany, where it occurs at several thin horizons with bedding planes crowded with the one species. Thus *Bositra,* like *Claraia* in the Lower Triassic, was apparently an opportunist that flourished in high-stress environments characterized by oxygen deficiency, and the *Bositra* layers mark times when conditions on the sea bed were ameliorated from anoxic to dysoxic.

The Jet Rock marks the establishment of euxinic conditions following a marked eustatic sea-level rise, one of the most pronounced in the whole Jurassic. Nowhere else in the world is a black shale so clearly exposed at this horizon. The existence of a carbon isotope record strongly suggests a global anoxic event leading to increased burial of organic matter on the sea floor. At the family and genus level no major extinction is evident, but at the species level it is striking and exactly coincident with the anoxic horizon; indeed, it marks the most important extinction event of global extent in the whole of the Jurassic. The great majority of benthic and nektobenthic species became extinct, at least in Europe, where the best stratigraphic sections are found, but planktonic organisms survived unscathed, because anoxia affected only the deeper waters of the epicontinental seas.

Cenomanian–Turonian boundary (mid-Cretaceous)

The Cenomanian and Turonian stages are widely recognized as marking one of the all-time highstands of global sea level, but details of the changes during the Cenomanian–Turonian boundary interval remain somewhat contentious. The stratigraphy and faunal successions of only two regions, north-west Europe and the Western Interior of North America, are known fully. A number of claims of an extinction event at this time have been made, based mainly on the foraminifera and molluscs. Facies that indicate at least dysoxic bottom conditions occur in a number of regions, including the deep ocean floor as known from boreholes; and the carbon isotope record is indicative of an oceanic anoxic event, which is most plausibly interpreted as marking increased burial of organic matter. Uncertainty has, however, persisted about the precise timing and importance of the claimed extinction event. Andy Gale and Andrew Smith of the Natural History Museum in London and their colleagues, who have made a careful study of the relevant sections in southern England, express doubts about the reality of the extinctions there.

Cambrian events

As mentioned in the last chapter, there was a mass extinction late in the early Cambrian that affected a variety of fauna, including reef-building archaeocyathids and trilobites. A. Y. Zhuravlev of the Moscow Palaeontology Institute has resolved the extinctions into two events. The earlier and more severe of the two, termed the *Sinsk Event*, is associated with an extensive spread of laminated black shale facies and a corresponding change in carbon isotope values. Since the black shales contain abundant monospecific assemblages of acritarchs (a group of phytoplankton), Zhuravlev has suggested that the extinctions were caused by an anoxic event, perhaps related to a phytoplankton bloom and resultant eutrophication, in which oxygen was used up because of the excessive productivity. Stratigraphic data are as yet, however, insufficient to determine whether or not the Sinsk Event was truly global.

By the late Cambrian the trilobites had attained their peak of both diversity and abundance, and they dominate nearly all fossil assemblages of this age. On the evidence of his studies in the western United States, the American palaeontologist Pete Palmer has distinguished a succession of trilobite faunas, separated by extinction horizons, that he has termed *biomeres*. Among the various contenders for possible extinction mechanisms, marine deepening or transgression, and associated anoxia appear to be the strongest, but the global character of these events has not yet been demonstrated convincingly.

Devonian–Carboniferous boundary

Although less important than the late Devonian mass extinction, the extinction event at the Devonian–Carboniferous (D–C) boundary was notable for the total disappearance of an important Devonian fish group, the placoderms, and the near-eradication of the ammonoids and two nektic groups; there were significant extinctions also among varied benthic groups. As the evidence comes from German stratigraphic sections, it has been termed the *Hangenberg event* (compare the Kellwasser event of F–F age). As for the Kellwasser event, anoxia is the leading candidate for the Hangenberg extinctions, and the evidence is essentially the same: global development of black shale facies, trace-metal signatures of intense anoxia, and the correspondence between the extinctions and the onset of anoxic deposition.

Palaeocene–Eocene boundary

Benthic foraminifera are the most abundant fossils of deep-water organisms, as known from the study of oceanic borehole cores. Although they were apparently little affected by the K–T boundary extinction event, they suffered a striking mass extinction at the end of the Palaeocene. According to the American micropalaeontologist Ellen Thomas about half the species became extinct. Nothing comparable is known at other horizons in the Cenozoic. The planktonic foraminifera were unaffected, in marked contrast to their fate during the K–T event. The Palaeocene–Eocene event was clearly global in extent, and the consensus among those specialists who have studied the borehole cores and the microfauna is that the extinction was caused by an episode of widespread oxygen deficiency in deep- and intermediate-water masses, leading to dysoxic rather than anoxic conditions. Using the best available timescale, derived from biochronological and palaeomagnetic data, Thomas has estimated that the event was geologically brief, lasting less than 25,000 years. Jim Kennett of the University of California at Santa Barbara estimates a duration of only 3000 years at the most.

We can conclude that sea-level changes seem to be strongly associated with many marine mass extinctions, but they are rises in sea level rather than falls, and are associated with the spread of anoxic or dysoxic bottom waters. Newell's idea that reduction of a habitat area would provoke an increase in extinction rate can still be applied, because the refugia with normally oxygenated bottom water would have been severely reduced during the more intense anoxic episodes, and in all likelihood no deep-sea regime would have been available (as it would at the present day). Anoxic events are, of course relevant only to marine life. To consider life on the continents as well as in the sea, we have to look at the role of climatic change.

7

Climate change

Unlike the other factors that have been invoked to account for mass extinctions, climate change is manifest to us all, whether we travel from the tropics to the poles or experience the seasonal cycle. Over a longer timescale, the issue of global warming in the recent past and likely future, and its probable consequences for other aspects of the environment, has occupied a considerable amount of media attention. Those people who are unaware of the likely consequences of the burning of fossil fuels cannot count themselves as well educated. Over a longer timescale, geologists have been aware for many decades of significant climatic changes on a global scale leading to the appearance and disappearance of polar ice caps on a number of occasions. Steve Stanley, the distinguished palaeobiologist at Johns Hopkins University in Baltimore, has actively promoted the view that episodes of climatic cooling are the most likely cause of mass extinctions. However, we must consider also the significance of global warming, and for the continents, at any rate, the possible effects of changes in the humidity–aridity spectrum. Before examining the relationships between climatic change and mass extinctions we need to examine the criteria from the stratigraphic record that geologists use to determine ancient climates, or palaeo-climates.

Criteria for determining palaeoclimate

The most obvious way of detecting cold conditions in the past is to find evidence of the presence of ice. At the present day the sedimentary deposits associated with glaciers and ice sheets, which occur where melting ice dumps its rock load, range in grain size from boulders and pebbles to finely ground rock flour. Such deposits are known as *boulder clay* or *till*, and ancient examples consolidated into resistant rock as *tillites*. The surfaces of hard rock that have underlain substantial ice sheets bear characteristic linear striations indicating the former direction of ice movement, such as glaciers moving up or down a U-shaped valley. The striations are produced by pebbles embedded in the ice, and are a unique marker for glacial action. In the 1830s Louis Agassiz, the great Swiss naturalist, extrapolated from his knowledge of the margins of Alpine glaciers to propose that the whole of northern Europe had been covered by one or more ice sheets in the recent geo-logical past. His research was based primarily on finding glacial striations in such areas as the uplands in Britain and northern Europe, together with extensive deposits of till in the lowlands. Agassiz's ice-age theory proved very controversial for several years, but eventually geologists were converted *en masse* by the overwhelming strength of the evidence. One recalls the mass conversion of Earth scientists to plate tectonics, with its implications of continental drift, in the late 1960s, but Agassiz lived to acknowledge the change, unlike Alfred Wegener, who died in 1930 when continental drift was still decidedly heretical. So overwhelmingly persuasive was the evidence in both northern Europe and northern North America that the notion of land-based ice was and remains universally accepted as fact. This raises an interesting question about the effective conversion of theory into fact. Philosophers no doubt insist on a distinction, but scientists operate in a more robustly empirical way. Since like everyone else they are stimulated by a challenge, they move on to other problems. Quaternary geologists still find plenty to argue about.

Till is not the only sedimentary rock indicative of ice. At the present day icebergs that have broken off the polar ice sheets and migrated to lower latitudes release boulders and pebbles as they melt. These large pieces of rock, termed *clasts*, may fall some distance through sea water until they become embedded in much finer-grained, uniformly muddy sediments. Ancient examples are known as *dropstones*; some contain marine fossils. Normal sedimentary processes involving the action of water and wind give rise neither to tills nor dropstones, although in a number of instances, as in mudslides moving down Andean valleys into the sea, rock-types that resemble glacially derived deposits can be produced. It is therefore necessary, where possible, to seek independent evidence to make the case conclusive, such as characteristic striations on the polished rock substratum, or at least on the clasts. Characteristic scratch marks on mineral grains can also be recognized on a minute scale, as revealed by the scanning electron microscope. Furthermore, if a purported tillite is laterally very extensive, with far-travelled clasts, the likelihood of its being produced by a localized mudslide is greatly reduced.

Purely lithological evidence for warm climates is not so straightforward. Substantial deposits of calcium carbonate are confined to low-latitude seas, and so the presence of thick limestone successions is generally taken as evidence of warm conditions. It must, however, be noted that limestones can also form from shells and shell fragments in quite cool conditions if the influx of sands, silts, and muds from the land is inhibited. The more decisive criteria derive from palaeontology and geo-chemistry, and we shall come to those shortly. Among deposits formed on land, coals and lignites, signifying former swamp deposits, indicate a substantial excess of precipitation over evaporation and therefore a humid climate, whereas evaporite deposits, such as salt and gypsum, indicate aridity. Deposits in semi-arid regions may contain characteristic calcareous bodies known as *caliche* or *calcrete*.

Fossil remains are an obvious source of information, but one that varies considerably in quality. Not only are some organisms more climatically sensitive than others, but their usefulness declines progressively with increasing geological age. Thus, for the Quaternary and even for the Neogene we can find fossils of species of animals and plants that are still living today, whose climatic tolerances can be carefully estab-lished. In the early Cenozoic, however, comparisons can be made at generic level at best, and in the Mesozoic the com-parisons can often be made only at family level. Furthermore, over long stretches of time running into tens of millions of years there exists the possibility that organisms may have changed their environmental tolerances somewhat. Back in the Palaeozoic the situation becomes even more difficult.

Despite such handicaps a great deal of valuable information can be obtained. For example, one of the most obvious features of present-day plant distributions is that, botanic gardens and land bathed by the Gulf Stream apart, palms are confined to low tropical and subtropical latitudes, because they cannot tolerate frost. The same is true for cycads and various types of fern. So when fossil examples are found at what in the geological past were high latitudes it strongly indicates a warmer global climate, as in the Mesozoic and early Cenozoic. The lack of evidence of polar ice caps at these times, in the form of glacial deposits such as tills, provides convincing support for this interpretation.

Among marine fossils, the best climatic indicators are reef-building corals, which are today confined to tropical latitudes. Some corals can live in cool waters in the deep sea, but true reef-builders depend on a symbiotic association with dino-flagellate algae, which means that the reefs are restricted to shallow waters within the photic zone. The corals that lived back in the Palaeozoic belonged to a different group from those of today, and this group did not survive the end-Permian extinction. If, however, extensive reef bodies occur within substantial limestone masses composed of these ancient corals together with other reef-builders, such as various types of calcareous sponge, it is probably safe to infer that they lived in warm water.

The third field of research that provides valuable palaeo-climatic information is oxygen isotope geochemistry. This is based on the research done a little over half a century ago by Harold Urey, who established that the ratio of the ^{18}O to the ^{16}O isotope varies with temperature. This is true both for sea water and for the organic shells secreted in isotopic equilibrium with it: the higher the ratio, or the 'heavier' the oxygen, the lower the temperature; and temperature can be determined by a well-known formula. As with all new fields of research, various problems have to be eliminated before trustworthy information can be obtained from fossils. The main problem concerns what is called diagenetic alteration of the sediment or rock, which takes place after its deposition and long after the sea water has disappeared. Interaction with meteoric groundwater on land, for example, substantially lowers the isotope ratio and thus can give palaeotemperature values that are too high. For this reason, the best results come from deep-sea drilling cores, where such alteration is minimal. Unfortunately, such substantially unaltered cores extend in time only back to the late Mesozoic.

Because of large global temperature variations in the Quaternary, in the ocean as well as on land, many pronounced oscillations of the oxygen isotope curve can be recognized in deep-sea cores. These oscillations have provided the basis for an invaluable isotope stratigraphy that makes possible fine correlation of events across the world. Further back in time, oxygen isotopes have provided the best quantitative data that we have of the pronounced global temperature fall that has taken place since the climatic optimum was reached in the early Eocene 55 million years ago. As shown in **Fig. 7.1**, there was a more or less steady fall in deep-sea water temperature through the Eocene, followed by a much sharper fall at the end of the epoch, when a time of relative stability ensued. After this time temperature continued to fall, apart from a rise in the early Miocene, but a further sharp fall in mid-Miocene times suggests the inception of a polar ice cap in Antarctica. Thereafter there was a strong fall to the Quaternary as the north

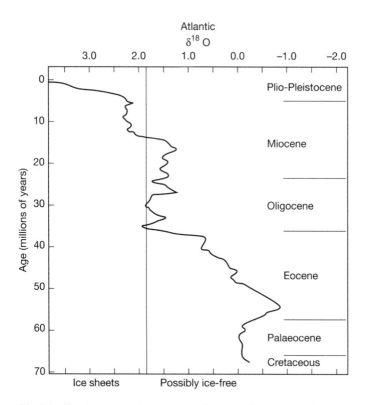

Fig. 7.1 Varying oxygen isotope ratio of foraminifera through the past 70 million years. Higher values of $\delta^{18}O$ signify lower temperatures.

polar ice cap developed. During a time interval of about 5 million years in the latest Palaeocene and early Eocene the oxygen isotope curve indicates that deep-ocean waters reached temperatures of up to 15 °C, as compared with about 4 °C today. Latitudinal temperature gradients in surface waters were low and warm-water pelagic marine organisms extended their geographic range into polar latitudes. Warmth-loving (thermophilic) vertebrates occurred in the Arctic. Crocodiles and monitor lizard fossils have thus been found in Ellesmere Island, Canada. The present-day tundra regions were occupied by forest, including broad-leaved deciduous trees. The soil types also suggest warm high latitudes, and clay-mineral associations in oceanic sediments indicate high humidity and intense chemical weathering in the Antarctic. Jack Wolfe, a palaeobotanist with the United States Geological Survey, has found empirically that the percentage of modern tree leaves whose margins are entire, i.e. non-serrated, gives a very reliable measure of mean annual temperature. Applying this principle to Palaeogene fossil leaves in Washington State he has established a curve of declining air temperature that is remarkably parallel to that found using deep-sea oxygen isotope data.

There is thus overwhelming evidence for a marked climatic decline from the early Eocene optimum onwards. The bearing of this on the theme of mass extinctions may be questioned, but it is directly relevant in one instance that will be discussed in due course. What I wish to emphasize here is that, of the various criteria for inferring temperature change, the only more or less reliable ones for short-term events that could be relevant to mass extinctions over geologically brief periods derive from oxygen isotope data.

Climatic changes in the Quaternary

Evidence of the sort outlined above has established beyond reasonable doubt that within the past two million years or so the Earth has been subjected to a succession of pronounced changes in temperature, with polar ice expanding in high latitudes and forests retreating in low latitudes during the cool phases and the reverse happening during the warm ones. If changes in air or water temperature were a significant factor in controlling organic extinctions it could reasonably be expected that there should have been an unusually high extinction rate at this time. It is perhaps surprising therefore that very few species extinctions have been recorded for the Quaternary, apart from those within the past few thousand years that are probably attributable to human activity.

I shall expand on this perhaps astonishing fact by recounting some of the research findings of my former Birmingham colleague Russell Coope, whose speciality is Quaternary beetles. It might be thought that, like other insects, beetles have only a low chance of preservation as fossils. This is not so, however, because in conditions favourable to the production of peats the elytra or wing cases of beetles have a high chance of preservation. According to Russell Coope, elytra are all you need to determine the species. When some entomologists scoffed, pointing out that one also needs to examine the genitalia, Russell Coope produced some genitalia superbly preserved as phosphate to support his identifications.

The standard method of determining temperature changes in continental environments has been to study fossil pollen from peats, because the environmental tolerances of the tree producers are well known. Russell Coope has found that his beetles are no less sensitive as indicators of air temperature than pollen. In warmer times both groups of organisms in the northern hemisphere have expanded their range northwards, and in cooler times they have retreated southwards. In other words, they have tracked their environments rather than becoming extinct. Evidently the area of refuge has been large enough for this phenomenon to occur again and again. Now insects are by far the most species-rich of all organic groups, and beetles are the most species-rich of all insect groups. For this reason alone one would have expected that beetles would speciate at the drop of a hat, at the merest hint of environmental disturbance. Not so, evidently. The great British evolutionary biologist J. B. S. Haldane, who was active in the first part of the twentieth century, was once asked by a pious inquisitor what he had inferred from his studies about the nature of the Creator. Haldane pondered for a moment and then replied, 'Well, he must have had an inordinate fondness for beetles.'

So why was there so little extinction in the Quaternary that could be directly attributable to temperature change? As we saw in Chapter 5, the same is true for Quaternary sea-level change. Perhaps the answer tentatively proposed in that case also has relevance in this one. The temperature change needs to be sustained beyond a critical time for recovery to be inhibited over a large enough geographical area to result in extinction. It is also possible that the organisms in question were relatively environment-tolerant, or *eurytopic*, having survived the less striking climatic vicissitudes of the Miocene and Pliocene as the Earth moved gradually into an Ice Age mode. Both these explanations could perhaps be applicable.

There is one interesting regional exception to the absence of substantial extinctions apart from those effected by our own species. According to Steve Stanley, in the western North Atlantic, off the east coast of the United States, two or more extinction pulses in the late Pliocene and Quaternary (between three and one million years ago) removed about two-thirds of the early Pliocene bivalve species. In marked contrast, there was a high survival rate among bivalve species on the northern Pacific borders, in California and Japan. Stanley attributes the extinctions to the association of climatic cooling and a particular palaeogeographic setting. According to Jeremy Jackson, formerly of the Smithsonian Tropical Institute in Panama, and now at the Scripps Institution of Oceanography in California, the mass-extinction event first documented by Stanley for bivalves north of the Caribbean occurred in a vastly greater region around the Caribbean and affected corals as well as molluscs. However, Jackson and his colleagues consider that changes in patterns of upwelling and nutrients may have been more important factors in causing the mass extinction than refrigeration due to the onset of the Quaternary glaciation, because there was no evident temperature decline in the southern Caribbean after the Pliocene and only a slight decline in Florida. A similar view has been expressed by an independent American palaeontologist, Warren Allmon, for gastropods of the *Turritella* group, which exhibited their highest rate of extinction in the western Atlantic in Pliocene–Pleistocene times, when almost all species became extinct. Allmon considers that nutrients were a more important factor than temperature. The gastropods could have been adversely affected by a collapse or a major reorganization of oceanic productivity associated with the formation of the Panama Isthmus, linking the North and South American continents and hence stopping flows of ocean water between the Pacific and Atlantic.

The relationship of climatic change to mass extinctions

Since the record of temperature change for the Cenozoic is the best we have, it makes sense to begin with that era. The time interval from the middle Eocene to the early Oligocene is now generally acknowledged to have been one of significantly increased extinction rates among both marine and terrestrial organisms. These extinction rates have generally been related to the pronounced climatic cooling that took place at that time (see Fig. 7.1). Claims of a mass extinction concentrated at the Eocene–Oligocene boundary have not, however, been supported by thorough analysis of the data.

Consider, for instance, the planktonic formaminifera. The analysis of this group from deep-sea cores by Gerta Keller and her colleagues provides the most comprehensive data set available for the study of marine faunal turnover. Among low-latitude taxa there was a major turnover from the late middle Eocene to the early late Oligocene, which involved more than 80 per cent of the species and took place more or less continuously over a period of about 14 million years. The overwhelming majority of species that became extinct at that time were surface dwellers that were replaced by more cold-tolerant forms that lived in the subsurface layer. This inference is based on oxygen isotope results, not morphology. Two intervals stand out as characterized by a brief but markedly intensified turnover: the middle–late Eocene boundary and the early–late

Oligocene boundary. Contrary to previous reports, there was no major faunal change across the Eocene–Oligocene boundary. A comparable picture is obtained from a study of the nanoplankton, mainly coccoliths, for which also we have an excellent record in deep-sea cores. More patchy data are available for benthic shallow-water groups such as molluscs, corals, and echinoderms, but the overall results are consistent with what is known from studies of the plankton. Many instances are known where genera that are confined to low latitudes today began a latitudinal contraction of their range from mid-Eocene times onwards.

The best fossil record from the continents derives from plants and mammals in Europe and North America. Palynological evidence from western Europe indicates that the Eocene–Oligocene transition was marked by the incoming of temperate elements, the loss of tropical and subtropical elements, and an increase in conifer pollen. Evidence of cooling starts in the late Eocene. In Germany and the Czech Republic there is macrofloral evidence of a change from a dominantly evergreen, subtropical flora in the late Eocene to a mixed evergreen and deciduous flora in the early Oligocene, indicating a warm but temperature-seasonal climate.

The rich record of Palaeogene mammals in the non-marine deposits of western North America enables us to make a more thorough examination of change through time than anywhere else in the world. The patterns of change in diversity are consistent with global cooling and the development of a less equable climate through the late Eocene and Oligocene. There was a concomitant shift from generally subtropical, closed-forest habitats to more open, savanna-like habitats in the Western Interior. The faunal turnover in the latter part of the

Eocene resulted in the extinction of many primitive mammal groups and the appearance of many modern families; there were high rates of both extinction and origination near the end of the Eocene. It is not clear from the present state of knowledge how stepwise or catastrophic the extinctions were. By the end of the Eocene more cursorial (running) adaptations appeared at a time coincident with the opening of new habitats. The changes in habitat that can be correlated with global cooling are likely to have been a major factor in causing the extinctions, rather than the temperature changes themselves. In Europe the most striking extinction event has been known for a long time as the *Grand Coupure*. Although traditionally placed at the Eocene–Oligocene boundary, it has now been redated as early Oligocene, about one to two million years younger than the epoch boundary. Approximately half the mammal genera became extinct within a million years, which was at least calamitous, if not catastrophic. In the course of this drastic change large-sized species became rare and medium-sized species disappeared. This, it is thought, indicates a change towards greater aridity and a more open environment, that is, from humid, warm-forested conditions to arid, colder, more open savannah-type environments. These changes appear to mirror those in North America, but the evidence of drying is much greater in America. Almost all the post-Grand Coupure taxa are immigrants from Asia or North America via Asia, which crossed the Bering land bridge. Thus, at least to some extent, the Grand Coupure was less a climatic event than a sudden influx of immigrants comparable to what is known as the Great American Biotic Interchange between North and South America in the late Neogene, which took place after the uplift of the Panama Isthmus.

Increased extinction rates among land organisms can be produced by increasing aridification in lower latitudes related to global cooling. This can be recognized in Africa, and especially in East Africa. Although mass extinction is not necessarily involved this topic deserves a brief mention because of its possible relevance to human origins. Humans, together with the apes, belong to the hominoid primates. Diverse tree-dwelling hominoid primates such as *Proconsul* and its relatives flourished in thick forests in the Miocene, but rapidly became extinct in the mid- and late Miocene, a time when forests were replaced by savannah as the climate became more arid. The emergence of bipedal hominids such as *Australopithecus* took place not long afterwards on a geological timescale. Unfortunately the lack of a hominid fossil record before about five million years ago, and of any fossil record for the African apes, is still a frustrating barrier to further understanding. Elizabeth Vrba, of Yale University, is the leading authority on fossil antelopes. On the basis of her study of this diverse group, she has proposed her *turnover pulse hypothesis*, in which she argues that extinction and origination have undergone coordinated sharp increases at restricted time intervals corresponding to times of sharp climatic change. Whether such punctuations within the past three million years are relevant to hominids, however, remains uncertain because of the sparsity of the fossil record.

Probably the most interesting event involving extinction is that at the boundary of the Palaeocene and Eocene. As mentioned in the previous chapter, about half the benthic foraminiferal species became extinct within a few thousand years, a phenomenon that ranks as a catastrophe for this group at least. **Figure 7.2** shows that this extinction corresponds exactly

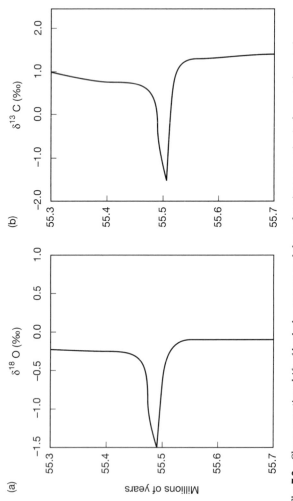

Fig. 7.2 Sharp negative shift of both the oxygen and the carbon isotope ratios in foraminifera at the Palaeocene–Eocene boundary. Isotope ratios are in parts per thousand (‰).

with a sharp and pronounced fall in both the oxygen isotope and carbon isotope ratios. The oxygen isotope data indicate a rise in temperature of several degrees. The pronounced reduction in carbon isotope levels is consistent with a marked fall in productivity, but there is apparently no decrease in the biogenic production of calcium carbonate ($CaCO_3$) at this time, which is in striking contrast to what happened at the K–T boundary. The extinctions of benthic species occurred before any large-scale changes in surface-water processes that could have affected primary production. Similarly, Ellen Thomas's research on the forams demonstrates an increase in the relative abundance of infaunal species, that is, deposit-feeding species that live within the sediment. This appears to rule out a decline in productivity as a causal factor.

What, then, could have caused such a major negative shift of the carbon isotope curve? The most plausible suggestion is that it is due to the release of methane from methane hydrates, because the carbon of methane (CH_4) is extremely 'light' in its isotope ratio. Methane hydrates are ice-like solids that occur in huge quantities beneath cold or deep ocean floors. Under conditions of either lowered pressure or increased temperature the solid hydrate disintegrates, releasing methane into sea water, and thereby lowering the carbon isotope ratio. Release and oxidation of methane in the large quantities implied by the isotope shift would have reduced oceanic oxygen levels, with a concomitant production of carbon dioxide. This could be the prime reason for the dysoxic event that is believed to have been the cause of the foram extinctions, as mentioned in Chapter 6. The trigger to the methane release is thought to be a pulse of increased sea-water temperature; both deep-sea and high-latitude temperatures increased by as much as 7 °C.

It is time now to turn our attention to events in pre-Cenozoic times, for which the climatic record is unfortunately much poorer. Some reasonable inferences can nevertheless be made by using various lines of evidence. We shall start with the biggest of the big five mass extinctions and shall conclude by mentioning an interesting lesser event.

1. End-Permian

Steve Stanley has argued that the end-Permian extinctions were the consequence of climatic cooling, which could also explain regression of the sea because of growth of polar ice caps. We have seen, however, that regression can be ruled out as a causal mechanism for the marine extinctions, and there is no evidence of substantial polar ice anywhere in the world after early mid-Permian times. To that time can be dated the youngest tillites and dropstones of the great Gondwana ice sheet that covered much of that supercontinent at various times in the Carboniferous and Permian. The oldest glacial deposits and striations occur in southern Africa and South America, and the youngest in India and Australia.

Much relevant information has come to light since Stanley made his proposal in the 1980s. The best evidence comes from the southern hemisphere, and is due to the Australian Greg Retallak, whose main speciality is fossil soils or palaeosols but who also has considerable palaeobotanical expertise. He has found that the latest Permian peats in Antarctica and Australia are comparable to those formed at similar high latitudes today where intense winter freezing occurs, but they are rapidly replaced by warm-temperature palaeosols in the earliest Triassic. In the slightly lower palaeolatitudes of the Karoo Basin in South Africa sedimentary evidence indicates a transition across the P–T boundary from a humid temperate to a hot semi-arid climate.

The floral evidence for temperature rise is even more overwhelming. The *Glossopteris* flora is highly characteristic of the Gondwanan supercontinent, and was used by Wegener as evidence for the former existence of that supercontinent, whose components today, South America, Africa, Australia, and India, are separated by wide stretches of ocean. Glossopterids are a very distinctive floral group confined to Gondwana in the Carboniferous and Permian. They occur in abundance in deposits interlayered with glacial deposits such as tillites and dropstones. It is therefore reasonable to infer that they were well adapted to cold climates in high southern palaeolatitudes. Retallak's work in Australia has established that the glossopterids suffered an abrupt, evidently catastrophic, extinction at the very end of the Permian and were replaced by a totally different flora in the Triassic. The most plausible interpretation is that a sharp rise of temperature drove them to extinction. Plant extinctions elsewhere in the world were less striking, but nevertheless more notable than earlier or later in the Permian and Triassic. The clearest global signature at the boundary has been discovered by palynologists and is known as the *fungal spike*. This marks an unusual proliferation of fungi, clearly indicating a massive increase in the production of dead trees. We can presume that a short, sharp shock was delivered to the world's vegetation.

Geochemical evidence for global warming comes from both oxygen and strontium isotopes. The only reasonably trustworthy oxygen isotope data have been obtained from a borehole in the Carnic Alps of Austria. The records show a sharp negative shift at the P–T boundary, indicative of a short-lived temperature rise of 6 °C. The strontium isotope measurements show that the $^{87}Sr/^{86}Sr$ trend during the Permian was one of long-term decline followed by an increasingly rapid rise through the latest Permian into the early Triassic. The consensus view is that the rapid rise in the strontium isotope ratio records a major increase in the global rate of continental weathering, a conclusion further supported by neodymium isotope data. There are two possible explanations: global marine regression increasing the land area, or enhanced chemical weathering triggered by increased humidity and atmospheric carbon dioxide levels. We ruled out the first alternative in Chapter 5. By far the most important geochemical signature of the P–T boundary comes from the study of carbon isotopes. A marked fall in the $^{13}C/^{12}C$ ratio is recorded globally, both in marine and continental strata, including palaeosols, and has been taken as evidence of a 'Strangelove Ocean' because of the drastic fall in productivity, which is comparable to the fall at the K–T boundary event. It is unclear, however, how a Strangelove Ocean could apply also to the continents. Taking into account the fascinating story now emerging for the Palaeocene–Eocene boundary, mentioned earlier in this chapter, we have to consider seriously the possibility that the isotope change is a consequence of release of methane into the ocean on a massive scale, lowering the isotope ratio in the sea water and ultimately in the atmosphere also, as methane was oxidized to produce carbon dioxide. As in the end-Palaeocene event, the trigger for the disintegration of methane hydrates in the bottom sediment would have been a sharp rise of temperature.

The best record of land vertebrate extinctions comes from the Karoo Basin in South Africa. Recent research by Roger

Smith and Peter Ward, published in 2001, suggests that a relatively sudden, possibly catastrophic, event, perhaps of 50,000 years duration or less, correlates with a mass extinction of terrestrial flora and fauna. Facies analysis of the sedimentary succession signifies a change from a wet floodplain to a dry floodplain environment, bringing about a faunal change to a recovery fauna composed of drought-tolerant organisms in the early Triassic.

2. End-Ordovician

Apart from the late Neogene globally and the Carboniferous and Permian of Gondwana, the late Ordovician is the only time in the Phanerozoic for which we have convincing evidence of polar ice caps, in the form of tillites and striated pavements, in the Sahara and adjacent regions. The end-Ordovician mass extinction would thus appear to be an obvious candidate for a cause related to global cooling. We saw earlier that this extinction turns out to have been a two-phased affair, and the second phase appears to relate to marine transgression and associated spread of anoxic waters (Chapter 6). Thus it is only with the earlier phase, starting in the final stage of the Ordovician, the Hirnantian, that we are concerned.

Indeed, the striking regression at this time, apparently global in extent, is generally related by Ordovician specialists to the growth of a south polar ice cap (Chapter 5). This raises the question of whether the increase in extinction rate among benthic faunas was due to loss of habitat (the Newell model), or cooling. Oxygen isotope data have been obtained from stratigraphic sections in both northern Europe and North America in an effort to resolve this question. The results show a pronounced positive shift of the oxygen isotope curve in the early part of the Hirnantian, which is indicative of a temperature fall in the palaeotropics, where these regions were located at the time. Unfortunately the large amount of shift that is recognized, of 4 parts per thousand, suggests a drop of 10 °C in the tropical ocean temperature. This fall seems to be implausibly high, suggesting almost inconceivably severe climatic changes in the latest Ordovician. As mentioned earlier, oxygen isotope results in the earlier part of the Phanerozoic are much less reliable than those for the Cenozoic and late Mesozoic, and so a large element of doubt must persist about the reliability of this result.

So far as decline in diversity is concerned, a cooling mechanism cannot account for the demise of many high-latitude, cool-water taxa, especially conodonts, trilobites, and graptolites, but it may still be applicable to the demise of lower-latitude forms. Generally, at times of glaciation, the warm tropical belt contracts but does not disappear. The survival of a distinctive fauna in a narrow tropical belt of limestones stretching from the ancient continents of Laurentia (North America) to Baltica (northern Europe) indicates that this was the case during the Hirnantian glaciation. Cooling is therefore unlikely to account for the loss of the numerous tropical taxa unless, of course, the oxygen isotope results (the $\delta^{18}O$ values, which express the changes in the $^{18}O/^{16}O$ ratio) are a true proxy for major equatorial cooling.

3. Frasnian–Famennian boundary

Paul Copper of the Lawrentian University in Sudbury, Ontario is the chief proponent of a model involving Frasnian cooling and extinction. In favour of this model he has put forward various lines of supporting evidence. That from the fossils includes the preferential elimination of reef taxa and other low-latitude inhabitants, and the proliferation of presumed cold-water groups during and after the crisis interval. The presence of late Devonian glacial sediments in Brazil and indications of rapid sea-level changes from stratal successions of this age suggest glacioeustatic control. The cooling is believed to be the consequence of a change in plate configuration that restricted the circulation of equatorial currents and deflected cool, high-latitude currents into the tropics. Of all the proposed extinction mechanisms this is the most slow-acting because it is a product of continental drift. Copper therefore emphasizes the gradual nature of extinctions extending through Frasnian time.

There are a number of problems with this interpretation. The glacial evidence from Brazil comes from rocks that are much too young: Famennian at the earliest. The faunal data are also open to alternative interpretation because the most successful surviving taxa have not only a cool-water tolerance but also a deep-water habitat. Their proliferation during the crisis interval could, therefore, equally well reflect the spread of dysaerobic, deep-water facies in association with the Kellwasser transgression, mentioned in the previous chapter. Rather than cooling, a phase of exceptional warming in the late Frasnian–early Famennian time interval can be proposed, on the basis of oxygen isotope data that indicates water temperatures apparently reaching 40 °C, which is well beyond the lethal limit for virtually all multicellular organisms. (This suggests death by cooking!) Such an interpretation seems highly implausible because a temperature of 40 °C should have caused 100 per cent extinction, which is manifestly not the case. As mentioned earlier, oxygen isotope data are highly susceptible to disturbance as a result of later modification by diagenesis in meteoric groundwaters, whose light $^{18}O/^{16}O$ ratios give false (high) palaeotemperatures.

4. End-Triassic

As mentioned earlier, polar ice caps were lacking in both the Triassic and Jurassic periods, and floral and faunal palaeo-geographic distributions indicate that warm conditions extended into much higher latitudes than today. The conventional view is that there was no significant temperature change across the Triassic–Jurassic (T–J) boundary, and this is supported, for sea-water temperatures at least, by trustworthy oxygen isotope results from the Northern Calcareous Alps of Austria. This view has, however, recently been challenged for air temperatures by some interesting palaeobotanical research by Jenny McElwain of the Field Museum of Natural History in Chicago and Dave Beerling of the University of Sheffield. Their research, published in 1999, was on the rich terrestrial plant beds straddling the T–J boundary in East Greenland and southern Sweden, which record a striking mass-extinction event across the boundary, with only a small percentage of plant species surviving. The plant cuticle with its stomatal pores represents an important interface between the plant and its surrounding environment, through which gas exchange can take place. Observations on living plants grown in controlled experimental environments suggest that there is an inverse relationship between the stomatal density and the carbon dioxide content of the environment. So a determination of stomatal density on fossil cuticular material should give a measure of the carbon dioxide content of the atmosphere. McElwain and Beerling found a marked decline in stomatal density and stomatal index, a more accurate measure based on density, from the Triassic to the Jurassic, especially in the East Greenland data, and they inferred from this a sharp rise in atmospheric carbon dioxide and hence temperature through the greenhouse effect.

This interesting finding is subject to a number of caveats. The fossils used are all long-extinct gymnosperms, with no close living relatives, and the experimental work was conducted on angiosperms. Stomatal work on Cenozoic angiosperms, with much closer living relatives, shows some promise, but there are numerous complications, because stomatal density on a given plant can vary with parameters other than temperature. Another problem is that in central Europe neither macroplants nor pollen and spores indicate a mass-extinction event, although pollen and spores in the northeastern United States apparently do indicate such an event. It would be interesting to learn what fossil cuticular material might reveal about stomatal density changes across the T–J boundary in central Europe. Clearly much more research needs to be done in this promising field before McElwain and Beerling's results can be accepted with more confidence.

5. End-Cretaceous

The environmental scenarios proposed for the consequences of bolide impact include both sharp rise and fall in temperature, but the timescale involved is much too brief to leave a record in oxygen isotope data from K–T boundary rocks across the world. On a longer timescale, the oxygen isotope evidence from deep-sea cores indicates a more or less progressive decline of sea-water temperatures from a mid-Cretaceous peak to the late Maastrichtian. The Maastrichtian data show strong fluctuations, with some indications at least locally of temperature rise just before the K–T boundary, but with a general trend towards cooler conditions accelerating in the mid-Maastrichtian. As mentioned above, the Maastrichtian marks an environmentally 'noisy' time and there is a strong likelihood that climatic change, allied for marine groups with oceanographic change, is heavily implicated in the increase in extinction rate observed in many organisms before the end of the period. Much more needs to be done, however, to tease out the various possibilities with more substantial and refined data.

In addition to the 'big five', some consideration also needs to be given to one of the lesser events, namely the early Toarcian mass extinction. As mentioned in the previous chapter, this extinction corresponds closely with a marine anoxic event. The positive shift in the carbon isotope curve is a clear indication of this event, as are the widespread black shales at this horizon, but in parts of Europe at least, where the best successions occur, this positive shift is preceded by an even more notable negative shift. Steve Hesselbo of Oxford University and his colleagues, working on English and Scandinavian rocks, found lower carbon isotope values at the same horizon also in fossil wood. The interpretation they offered was similar to that now generally accepted for the end-Palaeocene event, that the negative shift is the consequence of the release on a massive scale of methane from methane hydrate, with a consequent 'lightening' of the carbon isotope ratio. That the atmosphere was affected as well as the hydrosphere is supported by the isotopic data from fossil wood. The likely trigger for the methane release is thought to have been a pulse of increased temperature. Whereas the evidence in the form of an oxygen isotope shift for the Palaeocene event is very strong, that for the Toarcian event is not, because such few oxygen isotopic data as are available are likely to have been distorted by changes that took place after the sediments had consolidated (diagenesis), but this does not mean that there was no temperature rise as claimed. The interpretation given by Hesselbo and his colleagues seems the most plausible one available.

In conclusion, one further matter needs to be alluded to, which acts as a link between this and the previous chapter. The higher the temperature of sea water, the less oxygen it can hold. Warming events therefore tend to promote dysoxia and anoxia. It is thus no surprise to find widespread early Toarcian black shales. Of more general significance is the very widespread occurrence of black shales or equivalent facies across the Permian–Triassic boundary. This provides further support for a warming event at that time.

8

Volcanic activity

Of all the geological phenomena that make an impression on the general public, volcanic eruptions must be the most spectacular. The sight of incandescent lava being spewed into the air or creeping down hillsides is a regular standby of television documentaries dealing with the natural world, and the invocation by Old Testament writers of fire and brimstone in their portraits of hell indicates how vivid is the image of volcanic activity that has persisted throughout recorded history. How significant, though, are erupting or exploding volcanoes as killers?

Many years ago I paid a standard tourist visit to Pompeii and Herculaneum, the Roman towns destroyed by the eruption of Vesuvius in AD 79. I was duly impressed by the exhumed streets and buildings, but what caught my imagination most were the plaster casts from Pompeii of human beings found in recumbent or prone positions, or huddled against a wall, at the locations where they had been enveloped almost instantaneously by incandescent ash. These had been produced during the nineteenth-century excavations by a team of archaeologists led by an ingenious professor, who poured plaster of Paris down holes in the volcanic rock occupying the rooms of Roman villas. The rock was then removed. Good examples of these plaster casts can be seen in the Naples Museum, and picture postcards are freely available for tourists visiting the Bay of Naples area. They make disturbing *mementi mori* to send to your friends and relatives.

It is not known how many died as a result of the Vesuvius eruptions: perhaps not too many, as most inhabitants had time to escape. This was not the case, though, with the eruption in 1902 of Mont Pelée in Martinique, during which no fewer than 30,000 inhabitants of St. Pierre – 'the Paris of the West Indies' – were killed as what is called a *nuée ardente* smothered the town (**Fig. 8.1**). There were only two survivors. The term *nuée ardente* is normally translated as 'glowing cloud' or 'glowing avalanche', though some of the more frivolous students I have taught have preferred their own mistranslation: 'ardent nudes'. *Nuées ardentes* are quite common phenomena associated with the more explosive volcanoes, flowing down their flanks at 80 to 170 kilometres per hour. They take the form of density currents composed of an incandescent solid–liquid suspension that flows freely on and in a cushion of gas and compressed air.

The largest eruption in modern historical times was that of the Tambora volcano on the Indonesian island of Sumbawa, directly east of Bali and Lombok. The fullest report comes from Stamford Raffles, the founder of Singapore, who at that time was British Resident in Malaya and the East Indies. Eruptions continued for several months, from April to July, and ash was carried hundreds of miles, in sufficient quantities to darken the air, so that 'day became night'. Out of a population of 12,000 on the island, only 26 survived.

Fig. 8.1 An example of a *nuée ardente* photographed in Martinique.

Some of the highest mass mortalities may be an indirect consequence of volcanic activity. Perhaps the most celebrated example is also in Indonesia: the island of Krakatoa in the Sunda Strait. This eruption was the subject of a detailed study (the first of its kind) by the Royal Society of London. The eruption of Krakatoa in 1883 was a minor one compared with Tambora, but its effects were greater. It produced one of the loudest sounds in recorded history, being heard as far as 3000 miles away. Mortality as a direct result of the eruption was negligible, but no fewer than 30,000 people lost their lives in Java and Sumatra as a result of the wave, or tsunami, induced either by the collapse of the island structure or by the ejected material plummeting into the sea, which flooded the coastal regions of these islands. *Lahars* are mudflows of volcanic ash, which can cause extensive damage and loss of life if they flow down valleys towards lower ground. Thus lahars induced by the Kelut volcano, which exploded in Java in 1919, flowed down valleys radiating from the volcano and killed 5500 villagers. A modern example concerns the town of Almera in Colombia. In 1985 an eruption of a volcano 40 miles away induced melting of a glacier, which in turn led to a torrential mudflow that enveloped the town, killing some 20,000 people.

Another modern catastrophe is indirectly related to magmatic activity in a quite different way. In the Lake Nyos (Cameroon) disaster of 1986, about 1700 people died from asphyxiation when carbon dioxide, suddenly released from the mountain lake, flowed down the mountain slope towards their villages. Carbon dioxide of magmatic origin had been entering the lake water column through seepages at the bottom of the lake. Eventually the lake erupted, sending a fountain of gas–water mixture to a height of at least 120 metres and creating a water surge that washed up the southern shore to a height of about 25 metres. Work was subsequently put in hand to remove the gas from the lake in order to prevent a repeat of this tragedy.

On rare occasions volcanic activity may affect the course of civilization. In 1470 BC or thereabouts there was a catastrophic eruption of the Santorini volcano on the island of Thera in the Aegean Sea directly north of Crete. As a result, a large part of the island subsided beneath the sea. A series of eruptions appears to have taken place, and major ash layers dated at this time have been found both in Crete and in sediment cores derived from deep-sea drilling in the eastern Mediterranean. As for Krakatoa, the main destructive agent is likely to have been a tsunami induced by the collapse of the Santorini caldera. This may well have devastated coastal settlements in northern Crete. In what archaeologists call the late Minoan 1 B period something happened that brought the Minoan civilization to its knees, so to speak. Fleets were destroyed and most of the palaces and coastal settlements in the north and east abandoned, never to be reoccupied. The balance of power shifted from Crete to Greece, where Mycenaean civilization was on the rise. No plausible explanation had been given until the discovery and dating of the widespread layers of volcanic ash, and it is now generally accepted that the Santorini eruption was a prime agent, if not the prime agent, in the destruction of the Minoan civilization. It is also quite likely that the substantial disappearance of occupied land into the sea sowed the seed of the Atlantis legend, as voiced by Plato in his *Timaeus* published in about 350 BC.

> Now on the island of Atlantis there arose a great and marvellous power with kings ruling over all the island, as well as many other islands and parts of the continent . . . But at a later time there occurred violent earthquakes and floods and one terrible day and night came when your fighting force all at once vanished beneath the earth and the island of Atlantis in similar fashion disappeared beneath the sea.

All these events were manifestly devastating locally, but life elsewhere carried on much as normal. For volcanic activity to be a potential player in mass extinctions, it has to be shown to have global significance as a result of gases reaching into the stratosphere and affecting climate.

Climatic effects of volcanic eruptions

In New England, 1816 was called 'the year without a summer', with average temperatures in June 7 °F (4 °C) below normal. The consequence was serious crop failures. The situation was even worse in Europe; disastrous crop failures led in places to famine, and the price of grain shot up dramatically. This unusual climate has been widely attributed to the eruption of Tambora in the Sunda Arc in the previous year. It should, however, be pointed out that there were other cold years in the early nineteenth century for which no volcanic explanation is available. Dust and sulphur dioxide (SO_2) from volcanoes form a sort of sunscreen. Sulphur dioxide is actually a greenhouse gas and its initial effect is to cause warming. However, it quickly reacts with water to produce sulphate aerosols that backscatter and absorb the Sun's radiation. Such effects are localized unless gases are injected into the lower stratosphere. They are thence rapidly dispersed around the hemisphere. Global cooling is well recorded in historical times, but the effect is usually only for one to two years because of the rapid 'rain-out' of aerosols. Dust from volcanic ash is less important, because it is usually washed out much more quickly.

Besides sulphur dioxide, carbon dioxide is volumetrically the most important gas from volcanoes. Because of the greenhouse effect, it produces global warming and, significantly, it stays much longer in the atmosphere than sulphur dioxide (**Fig. 8.2**). The greenhouse effect results from the fact that, whereas incoming light from the sun is of short wavelength, the energy radiated from the Earth is of long wavelength, and

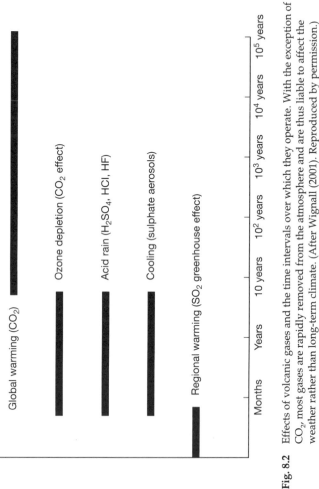

Fig. 8.2 Effects of volcanic gases and the time intervals over which they operate. With the exception of CO_2, most gases are rapidly removed from the atmosphere and are thus liable to affect the weather rather than long-term climate. (After Wignall (2001). Reproduced by permission.)

the more carbon dioxide there is in the atmosphere, the less transparent it is to outgoing radiation. Among the other volcanic gases, only chlorine (Cl) is postulated to occur in sufficient quantities to cause environmental harm as a result of local acid rain, in the form of hydrochloric acid, and ozone depletion. It is, however, rapidly removed from the atmosphere (Fig. 8.2). As is evident, carbon dioxide cannot be removed rapidly. The estimated yearly output of volcanic carbon dioxide is about 10^{11} kg, a figure dwarfed by the burning of fossil fuels, for which the equivalent figure is 10^{13} kg. Both values are minute in comparison with the amount of carbon in the atmosphere and oceans. Single eruptions are thus unlikely to cause any noticeable increase in this important greenhouse gas, but because of its long residence time the cumulative effect could be considerable.

The largest explosive eruption in the recent geological past was of the Toba volcano in Sumatra, 73,000 years ago. This left an elongated caldera about 50 km long and 100 kilometres across. Ash was spewed over much of the Indian Ocean, leaving a layer on the ocean floor up to 10 centimetres thick at a distance of 1300 miles. The eruption has left a clear record in a Greenland ice core, but oxygen isotope data from this core lend no support to any postulation of significant climatic change recorded in the seas of this time. This could well be bound up with the high thermal capacity, and hence thermal inertia, of the oceans. Even a fall in atmospheric temperature of 3–5 °C over a few years is unlikely to change sea-surface temperatures. The effect could, though, be more marked for terrestrial environments.

We have so far confined our attention to the explosive volcanoes that produce silica-rich lava and ash. In terms of the volume of material emitted, the most important volcanic activity is, however, from more or less non-explosive flood basalts, which are silica-poor, erupted from fissures. These form the bulk of what are called Large Igneous Provinces (LIPs for short), none of which has been formed in the past few million years (**Fig. 8.3**). Basalts are much richer in sulphur dioxide (SO_2) than the more silica-rich lavas of explosive volcanoes, and so the global cooling effect of the sulphate aerosols they produce is considerably greater. The only substantial example of flood-basalt fissure eruptions in modern times is that of Laki, in Iceland. Eight months of eruptions started in June 1783, with the ejection of enormous quantities of sulphur dioxide. It has been estimated that a hundred million tons of sulphuric acid rain descended over this period. This is approximately the same as the amount of acid rain that falls today on the whole Earth in an entire year. A significant cooling took place in the northern hemisphere; it was 7 °F cooler than average in eastern North America during the following winter, although more normal temperatures returned during the next two to three years. The European winter was especially harsh, and a bluish haze extended from Iceland as far as Siberia and North Africa. Acid rain destroyed most of Iceland's crops and 75 per cent of its livestock, and the effects reached Norway and England. That remarkable polymath Benjamin Franklin, serving at that time as American Ambassador in Paris, was the first person to suggest a relationship between volcanism and climatic cooling, on the basis of the Laki eruptions.

Since fissure eruptions are the most quiescent form of volcanic activity, it has been doubted whether they can inject gas into the stratosphere. Only the presence of what are called 'fire fountains' along the length of the fissures provides a potential

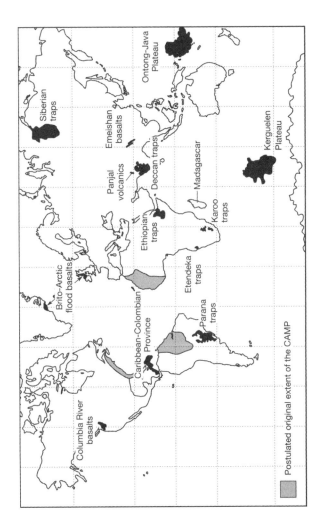

Fig. 8.3 Global distribution of continental flood-basalt provinces and oceanic plateaux. CAMP, Central Atlantic Magma Province. (After Wignall (2001). Reproduced with permission.)

mechanism, especially in high latitudes where the troposphere is thinnest.

We can sum up at this point by saying that volcanic activity in the recent past is likely to have affected the climate on a timescale ranging from a few months to several thousand years (Fig. 8.2). The shorter-term effects rank as weather rather than climate, and longer-term effects are not clearly demonstrated. This does not, however, take into account ancient Large Igneous Provinces, which exceed by many orders of magnitude the volcanic activity of historic times. Many LIPs have been shown to record brief bursts of activity over a period of about a million years, and recent improved radiometric dating generally shows a close correlation with mass extinctions. Indeed Vincent Courtillot, a geophysicist at the University of Paris (Denis Diderot) enthusiastically writes that the correlation is 'almost perfect'. Let us now explore the relationship in some detail.

Flood-basalt provinces and mass extinctions

Paul Wignall has undertaken a thorough analysis of the relationship between LIPs and mass extinctions, both major and minor. His map, reproduced here as Fig. 8.3, shows their global distribution. It must be borne in mind that on the Mercator projection used here areas in high latitudes are exaggerated with respect to those in lower latitudes.

Large Igneous Provinces, including Continental Flood Basalt Provinces (CFBPs) consist of a succession of subaerial and submarine sheet flows that range in volume from several hundred to several thousand cubic kilometres. Many were erupted in geologically brief intervals of one to two million years; the duration of the peak eruptive interval may be as short as 10,000 years. LIPs are widely regarded as the product of what are called *mantle plumes* that have ascended from deep within the mantle, possibly quite close to the core–mantle boundary, before impinging on the base of the lithosphere and subsequently producing eruptions of basaltic lava. Melted igneous rock is produced by decompression and consequent rifting of the lithosphere, the relatively brittle outer layer of the Earth. Melt volumes are greatest when the lithosphere is thin and able to extend freely, as at oceanic sites. The initial arrival of the plume at the base of the lithosphere is postulated to cause widespread doming, with elevations of one to four kilometres. It has been proposed by a few research workers that LIPs can be generated by bolide impact. This is generally considered unlikely because impact excavation, however large, is incapable of generating the vast volumes of lava found in flood-basalt provinces.

In considering the major episodes of mass extinction we have to take into account the fact that the oldest LIP recognized is no older than Permian in age. We therefore pass over both the end-Ordovician and the late Devonian events without comment.

1. End-Permian

The flood basalts of what are known as the Siberian Traps cover only a modest area of 340,000 square kilometres in north-west Siberia, but the inclusion of other associated igneous rocks greatly increases the total area to 1.5 million square kilometres, and the total volume has been estimated as at least 1–2 million cubic kilometres. (The curious term 'trap' is derived from the old Swedish word for staircase. If you visit a district where a succession of lava flows is well displayed on a slope, as, for example, the area adjacent to the east coast in northern Skye, you will quickly understand why the term has been so applied.) Recent exploration of deeply buried contemporary basalts encountered in boreholes in the West Siberian Basin adjacent to the exposed area to the east has established that the area of the flood basalts has at least to be doubled. The Siberian Traps province is thus even more significant than hitherto believed. Radiometric dating indicates that the basaltic eruptions began at about the time of the Permian–Triassic boundary, and that they persisted for about a million years or less. Volcanic ash is unusually abundant for a CFBP. Much of our present knowledge is derived from a region that represents less than 10 per cent of the entire province; some surprises may possibly be in store once we know more about the more poorly exposed areas. A major problem in interpreting the Siberian Traps is their setting within the heart of an ancient continent, which makes them in this respect unlike virtually all the other LIPs, which occur at locations marked by subsequent continental rifting. Another difference is the absence of any evidence of significant pre-eruption doming. A mantle plume nevertheless remains the preferred model.

In view of what has been said earlier in this chapter, the most obvious environmental effect of the Siberian Traps eruptions would have been global warming as a result of the expulsion into the atmosphere of significant quantities of carbon dioxide. This would probably have had two effects on the ocean system: first, a decline in the equator-to-poles temperature gradient; secondly, an increase in the tendency towards anoxia because of the lower solubility of oxygen in warmer water and the greater sluggishness of water circulation. As noted earlier, one of the most striking features of the Permian–Triassic boundary throughout the world is the pronounced negative shift of the carbon isotope ratio. A collapse in biological productivity is unlikely to produce more than a small isotope shift, and Doug Erwin of the Smithsonian Institution in Washington D.C. was the first to suggest breakdown of methane hydrate as the likeliest cause. There may be as much as 10^{19} grams of carbon locked up at present in methane hydrate beneath modern seas. If only 10 per cent of this were released to the atmosphere as methane, or converted to the other significant greenhouse gas, carbon dioxide, it would be sufficient to cause the observed carbon isotope ($\delta^{13}C$) shift because the carbon in methane is extremely 'light', that is, it has an unusually high ratio of ^{12}C to ^{13}C. The volcano could have acted as a trigger to release the methane, which could have given rise to a 'runaway greenhouse' because of the positive feedback. This could have been checked if the increased warmth had stimulated an increase in plant activity, which would result in an increased removal of carbon dioxide from the atmosphere by photosynthesis.

2. End-Triassic

The area of exposed flood basalts for the end-Triassic event is rather limited, and they are best seen in the eastern United States (Fig. 8.3). Associated igneous rocks representing shallow intrusions into the crust, such as sills and dykes (as distinct from lavas extruded at the surface) can, however, also be taken into account here. These intrusions have been found in such widely scattered regions as French Guyana, Surinam,

Fig. 8.4 The Columbia River of the American Northwest traverses the Columbia Plateau lavas of Miocene age, an example of a flood basalt province.

Guinea, and Brazil, and when these areas are taken into account it has been estimated that the area of the LIP could have been as much as seven million square kilometres, with a volume of more than two million cubic kilometres. Because of its location, on the periphery of the central part of the Atlantic, this LIP is called the Central Atlantic Magmatic Province, or CAMP for short (Figure 8.3). The best dating is from the Newark Basin basalts in the north-eastern United States. This gives an age of about 200 million years, almost exactly coincident with the T–J boundary. Dates from other parts of the province cluster around this figure, indicating a pulse of volcanic activity at about, or more likely shortly after, the end of the Triassic. The location of the province clearly relates to an early phase of tensional tectonics before the opening of the earliest part of the Atlantic early in the mid-Jurassic, which marks the start of the break-up of Wegener's supercontinent Pangaea. The Central Atlantic Magmatic Province supplies the strongest evidence of doming prior to the volcanicity, and thus of the rise of a mantle plume. This is the likeliest cause of the extensive end-Triassic regression on both sides of the present Atlantic, which was quickly followed by the earliest Jurassic transgression, mentioned in Chapter 5.

What other environmental consequences might there have been? An episode of climatic warming has been inferred by palaeobotanists on the basis of their stomatal density studies in Greenland and southern Sweden, and CAMP volcanism has been invoked to account for this. However, as mentioned in Chapter 7, their fascinating work is still somewhat controversial, and there is no clear evidence yet of climatic change on a global scale. Within the past few years carbon isotope studies of T–J boundary sections in Europe and North America have established the presence of a negative shift at the boundary comparable to, but much smaller than, that at the P–T boundary. This shift is consistent with a short-term warming event if it was due to the release of methane from methane hydrates in the sea bed.

3. End-Cretaceous

The Deccan plateau of western India is the site of another huge continental flood basalt province, occupying an area of half a million square kilometres, with an original volume generally estimated as two million cubic kilometres. The maximum thickness of 2.5 kilometres is in the Western Ghats. Little-known offshore extensions suggest that the volume might be much greater. Both magnetostratigraphic and radiometric dating indicate that the duration of the basaltic eruptions was about a million years, straddling the K–T boundary. Some research suggests a greater length of time, but the main peak of eruption was probably at the boundary. If we assume a million-year eruptive phase, modelling indicates a global temperature increase of less than 1 °C. Most researchers have therefore regarded volcanic sulphur dioxide as having caused the most environmental damage. However, the Chicxulub impact in Mexico is thought to have expelled no less than 10^{18} grams of sulphate into the atmosphere; so in this respect bolide impact would have been far more significant. Some research workers have favoured bolide impact as a trigger for the Deccan Traps eruptions, but the evidence clearly indicates that they started before the time of impact.

We now turn our attention to other LIPs (Fig. 8.3) and their possible relationship to lesser mass-extinction events. They will also be dealt with in age sequence.

4. Panjal Volcanics and Emeishan

The Panjal volcanism of north-west India and the Emeishan basalts of south-west China were probably erupted over a period of less than a million years at around the Middle–Upper Permian boundary, and they represent the oldest LIPs that are known. They may be related to the important extinction phase recognized at this time in China, which has been attributed to marine regression (see Chapter 5). This could signify regional doming prior to the Emeishan eruptions as a consequence of the rise of a mantle plume.

5. Karoo and Ferrar Traps

The Karoo basalts and associated igneous rocks of South Africa and the Ferrar of Antarctica form part of what was formerly a continuous province and mark the opening stage of the break-up of the southern supercontinent Gondwana. Together they represent over two-and-a-half million cubic kilometres of lava, erupted during a brief interval about 183 million years ago. This corresponds in time to the early Toarcian marine mass extinction of the early Jurassic. The pronounced negative shift of the carbon isotope curve mentioned in Chapter 6 can be related to these southern hemisphere eruptions, with the volcanism triggering the release of methane from methane hydrates and an ensuing pulse of global warming. As indicated in Chapter 6, anoxia is the evident cause of extinction, and an increasing tendency towards anoxia would have been favoured by the temperature increase, as for the end-Permian extinction event.

6. Paraná and Etendeta

The Paraná basalts of southern Brazil and the equivalent Etendeta basalts of Namibia were originally part of an extensive province one-and-a-half million square kilometres in area, which became separated with the opening of the South Atlantic. They were erupted over a time interval of only about six hundred thousand years in the earliest Cretaceous (Valanginian or Hauterivian stages), a time of low extinction rates.

7. Ontong Java Plateau

The Ontong Java Plateau is the largest single volcanic province on Earth. It occupies an area of two million square kilometres and forms a huge swell on the floor of the south-west Pacific. It is dated as late Barremian (early Cretaceous) and is not correlated in time with any mass extinction.

8. Caribbean–Colombia Plateau and Madagascan flood basalts

The Caribbean–Colombia Plateau and Madagascar are widely separated but the basalts of both igneous provinces are of the same age. The dates obtained for the lavas are mid-Cretaceous (Turonian), and they are thus slightly younger than the mass extinction at the Cenomanian–Turonian boundary. There was a reversal of a Cretaceous warming trend in the early Turonian. A possible environmental scenario is that warm and humid conditions enhanced the input of nutrients into the ocean, increasing primary productivity and thus increasing the burial of organic carbon – effectively a negative feedback.

9. Brito–Arctic flood basalts and the North Atlantic Igneous Province

Estimates of the volume of the basalts of the Brito–Arctic and North Atlantic igneous provinces range from two million cubic kilometres to six million cubic kilometres if offshore flows are included. The earliest eruptions took place about 61 million years ago, but the peak of the volcanic activity was about 56 million years ago, a date slightly earlier than the Palaeocene–Eocene boundary. The volcanism could therefore very probably be the chief factor responsible for the end-Palaeocene mass-extinction event in the deep sea, which was clearly associated with an episode of warming and dysoxic conditions. The release of methane from hydrates is required to explain the pronounced negative shift in the carbon isotope curve.

10. Ethiopian and Columbia River basalts

The Ethiopian and Columbia River basalts (**Fig. 8.4**) are the youngest and smallest well-dated LIPs. The Ethiopian basalts are dated at 28–31 million years (Oligocene). The Columbia River basalts of the north-western United States are mid-Miocene. Neither time corresponds to a mass extinction.

The other LIP shown in Fig. 8.3 is the Kerguelen Plateau. This is an extensive submarine plateau close to Antarctica, comparable to the Ontong–Java Plateau, and exposed at the surface as the island of Kerguelen. The age of this LIP is uncertain and it cannot therefore be correlated at present with any mass-extinction event.

Conclusions

Of the ten well-dated episodes of major flood-basalt eruptions or LIPs, three are closely correlated with major mass extinctions (end-Permian, end-Triassic, and end-Cretaceous) and

four with minor ones (late Permian, early Jurassic, mid Cretaceous, and early Palaeogene). There is also a strong correlation
with episodes of oceanic anoxia or dysoxia, although in only
four instances is anoxia regarded as the proximate cause of
the marine extinctions (end-Permian, early Toarcian, end-
Cenomanian, and late Palaeocene).

So although Courtillot was over-optimistic in calling the
correlation between LIPs and mass-extinction episodes 'almost
perfect' it is indeed good. The correlation is best for the four
events in the middle part of the Phanerozoic, from the Emeishan to the Karroo–Ferrar. On the evidence of the radiometric
dates, the onset of the eruptions or the interval preceding them
is associated with the most damaging environmental change.
One such change may be widespread uplift. The links between
volcanism and extinctions after the Jurassic appear to be much
more tenuous. There appears to be no simple correlation
between the volume of the eruptive rocks and the intensity of
extinctions. The rapidity of the eruptions may possibly be a
more significant factor. The failure to find evidence for cooling
events triggered by volcanism may be due to the short residence time of sulphate aerosols in the atmosphere (Fig. 8.2).
Episodes of global warming, on the other hand, show a good
correlation, which is best for the end-Permian, mid-Cretaceous,
and early Palaeogene events, but less certain for those of the
end-Triassic and early Toarcian. If the eruption of the Deccan
Traps had little to do with the end-Cretaceous mass extinction,
as is argued here, it is a remarkable coincidence in time. The
lack of LIPs older than the Permian is puzzling. It could perhaps
relate to the plate-tectonic regime in the earlier Phanerozoic,
with a lack of major rifting episodes and continents tending to
converge on each other rather than split apart.

We can now integrate the knowledge acquired from reviewing the various likely causes of mass extinctions to see if any general pattern begins to emerge, and we can also consider the subsequent recovery of the organic world after the extinction episodes.

9

Pulling the strands together

In drawing together the various strands we first need to ask how catastrophic, as opposed to merely calamitous, the various mass-extinction events were. As was indicated in Chapter 3, there is no way in which the stratigraphic record can ever provide dates that are precise to within less than a few thousand years. Thus, the connection between a bolide impact and a catastrophic phase of extinction lasting no longer than a few years could never be established with a high degree of confidence from the record of the strata alone. All that can be done is to establish a pattern that is consistent with such a scenario. As was also pointed out in Chapter 3, a change that is drastic enough over an interval of a few thousand to a few tens of thousands of years can reasonably be described as catastrophic in the context of normal patterns of geological change extending over millions of years. Several events seem to qualify unequivocally: the end-Permian, the end-Cretaceous, and, on a smaller scale, the end-Palaeocene, which affected only one group of deep-sea organisms. It needs to be added, though, that the end-Cretaceous event seems to have been the culmination of a phase of increased extinction rates among a wide variety of organisms. Such patterns of catastrophic change cannot yet be ruled out for the other mass-extinction events, but decisive evidence is not yet forthcoming. A more gradual or multiple pattern of extinctions appears to be more likely for the end-Ordovician, late Devonian, and end-Triassic extinctions and also for more minor ones such as those in the early Jurassic and mid-Cretaceous. Catastrophic *coups de grâce* are quite possible, if not probable, as culminating factors for some of these events, but more detailed collecting and statistical work across the world is required to put forward a stronger case than has been made so far.

It has been claimed that the 'big five' mass extinctions are something special, as opposed to lesser extinction events, in so far as they were too drastic and rapid in their effects on many organisms to give time for normal Darwinian adaptive responses to operate. On the other hand many others, myself included, regard the 'big five' as merely the end-members of a whole spectrum of greater and lesser extinction events, most of which have not even received a mention in this book. In Raup's most recently expressed view (his 1994 paper), the 'big five' stand out because of their scale, but actually represent the right-hand tail of a smooth statistical distribution. In Jonathan Swift's words,

> So, naturalists observe, a flea
> Hath smaller fleas that on him prey;
> And these have smaller fleas to bite 'em,
> And so proceed *ad infinitum*.

As has been mentioned in Chapter 4, Sheehan and Hansen have put forward an interesting hypothesis to explain selective extinctions at the K–T boundary associated with impact, but we must now consider what patterns can be discerned for mass extinctions in general.

One reasonable expectation is that organisms with wide environmental tolerances (eurytopic organisms) should be more resistant to extinction than those with narrow environmental tolerances (stenotopic organisms). The problem in finding evidence for this expectation in the stratigraphic record is the danger of circular reasoning: those organisms that survived a given extinction event must have been relatively eurytopic! It is obviously better to seek some distinctive features of those organisms that were more extinction-resistant or extinction-vulnerable. On the basis of his comprehensive studies of late Cretaceous gastropod and bivalve molluscs Dave Jablonski has argued that wider geographic distribution prior to the K–T mass extinction conferred greater resistance to it. This seems a reasonable enough conclusion, but whether or not it is generally true for the Phanerozoic as a whole has not yet been convincingly demonstrated; Jablonski's findings have indeed been challenged by some research workers.

Another generally held belief is that organisms with a tropical distribution are more vulnerable to extinction than those occupying higher latitudes. This seems to be true of reef-building organisms, both corals and calcareous sponges, from early Cambrian times onwards. However, Raup and his student George Boyajian have demonstrated that extinction rates for reef taxa (the quintessentially tropical taxa) are not proportionally elevated relative to other so-called 'level bottom' taxa during mass-extinction intervals; this argues against latitudinal bias. Furthermore, Raup and Jablonski undertook an analysis of end-Cretaceous bivalves from this point of view. Only one major group of tropical bivalves, the rudists, disappeared at the K–T boundary (just before, in fact), but no other relationship between palaeolatitude and extinction could be recognized.

Perhaps the most easily measured features of a fossil taxon is the maximum size of its individual members, and fortunately this may have considerable biological significance. To state that larger organisms of a given species, genus, or family are more vulnerable to extinction seems to be one of the best generalizations that can be made at present. It has been reckoned, for instance, that no animal of weight greater than 25 kilograms survived the K–T extinction. That larger animals should be more vulnerable to extinction makes good sense. Larger animals are more likely to be rarer and breed less frequently. Both these factors will in all probability reduce their chances of surviving an environmental crisis.

One of the better generalizations in palaeobiology is that of phyletic size increase through time, or Cope's Rule, named after the distinguished American vertebrate palaeontologist Edward Drinker Cope, who competed actively in the late nineteenth century with a rival, O. C. Marsh, in hunting for dinosaurs and large Tertiary mammals in the American West. Throughout the animal kingdom, from minute foraminifera to huge mammals, there is a tendency for the maximum size of the individual organisms in a given taxon to increase more or less gradually through time. This does not appear to be matched by an equivalent gradual tendency to size decrease. This does not mean that evolution is always towards larger size; if that were the case, the world would be almost exclusively inhabited by giants. Clearly there are likely to be many adaptive advantages for some organisms in becoming smaller, but the change would appear to have been relatively sudden, and cannot be recognized as a whole sequence of progressively smaller fossils through a succession of strata.

The changes in size may relate to differing adaptive strategies. Although like all biological generalizations it has been challenged, one of the more useful ones that has resisted total dismissal is that of what are called r and K adaptive strategies. The letter 'r' is used to denote the reproductive rate of a given organism; 'K' signifies the carrying capacity of the environment. An animal with a high r will reproduce at a relatively rapid rate and mature early, attaining, in consequence of a reduction in its growth rate with adulthood, a maximum size of only modest dimensions. On the other hand, an animal with a high K is likely to reproduce more slowly, mature later, and attain a greater maximum size. Small size and high reproductive rate are characteristics of organisms living in a stressed environment which have an opportunistic lifestyle. They are adapted to take full advantage of limited opportunities for survival, whether of food resources or some other aspect of a fluctuating environment. What are termed 'K-selected organisms', on the other hand, can depend on a more stable environment and can therefore invest in quality rather than quantity in their reproductive strategy.

This pattern of change should relate well to Ager's 'long periods of boredom interrupted by moments of terror'. The moments of terror are the drastic changes in the environment which lead to mass extinctions, both major and minor. It follows that the organisms surviving these events should tend to be of small size for given taxa. This is indeed supported by what is normally found in faunas that lived in the immediate aftermath of extinction events. The phenomenon has been graced by a name, the *Lilliput effect*. One can accordingly propose a model. As stable conditions are restored, there is an adaptive premium on living longer and growing larger. But there is a natural built-in check on this process. Making the reasonable assumption that food resources are approximately constant, as organisms of a given species or genus grow larger they must perforce become rarer, so increasing their vulnerability to extinction during the next 'moment of terror'.

I put this model to the test some years ago by comparing two groups of abundant Jurassic molluscs, the ammonites and bivalves. These two groups had very different evolutionary characteristics. The ammonites both radiated and went extinct at a much more rapid rate than the bivalves, giving a much higher rate of taxonomic turnover through time, which makes them much more valuable as biostratigraphic indicators. To make the point more vividly, it is not too fanciful to compare these molluscs with two types of horses. The ammonites were the thoroughbred racehorses, dashing and exciting, but temperamental creatures highly vulnerable to injury or disease, whereas the bivalves were the shire-type carthorses, more stolid and plodding but very trustworthy, environmentally more resistant dobbins. The principal result of my study was to find that phyletic size increase among the faster-evolving ammonites took place at a correspondingly more rapid rate. I consider that this provides at least gratifying support for my model. It follows from this study that, since ammonites evidently became extinct 'at the drop of a hat', so to speak, being vulnerable even to minor environmental disturbances, they are less valuable to us in our endeavour to discern the more significant disturbances than the more stolid bivalves. If many bivalves became extinct at a particular stratigraphic horizon, that must reflect that something really big happened. What is true for bivalves is likely to have been true for many other benthic invertebrates, because ammonites are exceptional in having such a high rate of evolutionary turnover.

After mass extinctions, the earliest phases of recovery tend, indeed, to be dominated by geographically widespread groups of small-sized organisms that occur in low diversity but high individual abundance. In consequence, one or a few species will totally dominate the fauna or flora and can occur in huge quantities. These are sometimes called *disaster taxa* and are thought to have had an opportunistic lifestyle. The end of the recovery phase commonly coincides with the reappearance of organisms that seemed to have disappeared at the extinction horizon, the *Lazarus taxa*. This must signify that these taxa had found some undiscovered refuge in which to survive until environmental conditions improved again, or that they became so rare after the extinction event that their apparent absence in the directly overlying strata is due to our failure to find fossils.

A brief survey of recovery phases after the major extinction events will give some idea of what evidently happened. After the loss of the archaeocyathan sponges (a group with calcareous conical skeletons) following the major mass extinction that occurred late in early Cambrian times, metazoan reefs did not rebound until some 25 million years later, in the earliest Ordovician of China, with the appearance of bryozoans (a group of colonial organisms) and two groups of sponges (the stromatoporoids and lithistids, with calcareous and non-calcareous skeletons respectively). In contrast, non-reef marine ecosystems quickly rebounded well before the end of the Cambrian period.

Between the two phases of the end-Ordovician mass extinction the dominant representatives of the low-diversity marine faunas were geographically widespread, small-sized brachiopods, which are thought to have been environmentally tolerant opportunists. With the subsequent rise of sea level in early Silurian times there was an increase in taxonomic diversification, but little biological innovation is evident at this time. Lazarus taxa among the brachiopods, trilobites, and cystoid echinoderms (which were the dominant echinoderms in the Cambrian and Ordovician) have been recognized in Middle Silurian faunas following the Lower Silurian, which was relatively depauperate (i.e. low in diversity). In the case of the late Devonian mass extinctions, rugose corals and stromatoporoid reefs disappeared dramatically, and reefs did not reappear until well into the Carboniferous. Nearshore marine environments seem to have suffered fewer extinctions than those farther from the shore, and were a source of the eurytopic survival faunas of the final Devonian stage, the Famennian.

After the end-Permian mass extinction, most of the early Triassic faunas, both in the sea and on the land, were of low diversity, and full recovery started no earlier than the latest, Spathian, stage. In marine environments, the 'paper pecten' or 'flat clam' *Claraia*, occurring in huge quantities in monospecific assemblages, was evidently an opportunist tolerant of oxygen-deficient environments. Locally, stromatolites reappeared in fully marine settings. These structures, produced by cyanobacteria, had been widespread in the seas of late Precambrian times, but had effectively disappeared in the Palaeozoic, probably because the grazing activity of the newly evolved metazoan benthos had driven them into marginal marine environments such as the intertidal zone. Today they are excellently displayed, for instance, in Shark Bay in Western Australia. The earliest Triassic seas were so deprived of benthic invertebrates that for a short time stromatolites could once more become well established. Terrestrial plant life was characterized by floras of low diversity dominated by the weedy lycopsid *Isoetes*, and arboreal floras were not restored until the Middle Triassic, at which time marine reefs also began to reappear, about six million years after the end of the Permian. (The reefs were composed of calcareous sponges belonging to a group, the Sphinctozoa, that consisted essentially of Palaeozoic survivors, and Scleractinian corals, a new group that had evolved after the rugose corals of the Palaeozoic died out at the end of that era.) There is thus a post-extinction *coal gap* as well as a *reef gap* in the stratigraphic record. Since the silica-skeletoned radiolaria also underwent mass extinction, there is a corresponding *chert gap*, because chert is formed by the reprecipitation of silica after dissolution of siliceous organisms, such as radiolaria, diatoms, or siliceous sponges. Early Triassic land reptiles were also depauperate, being dominated by the cosmopolitan *Lystrosaurus*. It thus took at least six million years before the world's biota began to recover fully from the end-Permian catastrophe and underwent the most complete ecological reorganization since the Ordovician radiation.

The end-Triassic extinction, in contrast, does not seem to have caused any drastic reorganization of ecosystems, except that once again reefs were drastically reduced, and the sphinctozoan sponges finally disappeared. The marine faunas of the earliest Jurassic, Hettangian, stage were limited in diversity, most organisms being small in size for their taxa, but by about two stages later, about 15 million years after the end of the Triassic, diversities were substantially restored and extensive reefs had reappeared.

With regard to the end-Cretaceous extinction, the earliest stage of the Tertiary, the Danian, is characterized by a survival interval at the start which is dominated by a single species of the planktonic foraminifer *Guembelitria*, a small, opportunistic form that rapidly expanded its geographic range in the absence of competition and quickly diversified. All subsequent planktonic foraminiferans are descended from this species and a second surviving species in another genus. By one to two million years after the mass extinction this group had begun to make a full recovery. The most detailed record of temporal change for land organisms comes from the plants. Several centimetres of basal Tertiary strata in North America are dominated by fern spores, indicating a classic type of opportunistic regeneration following severe ecological disruption. During 1.5 million years after the mass extinction angiosperm-dominated assemblages had undergone a full recovery, but this had probably begun only a few hundred thousand years after the extinction event. The deep-sea assemblage of benthic foraminifera that collapsed during the end-Palaeocene extinction event was succeeded for 100,000 to 200,000 years by a survival assemblage characterized by taxa that could survive low oxygen conditions. This assemblage was in turn succeeded by a recovery interval dominated by the migration of opportunistic taxa from shallow waters.

Extinction periodicity or episodicity?

Mention has already been made (in Chapter 3) of Jack Sepkoski's huge compendium of Phanerozoic marine families,
which was the basis for the recognition with his colleague
Dave Raup of the 'big five' mass-extinction events. Later on,
using a combination of methods involving Fourier analysis
and Monte Carlo simulation, they discovered to their great
excitement a statistically highly significant 26 million-year
periodicity in extinction events during the past 250 million
years. This was a most surprising result, because rare events
such as floods and hurricanes are apparently randomly distributed in time. In their paper in the *Proceedings of the National
Academy of Sciences* published early in 1984 they concluded
that 'it seems inescapable that the post-late Permian extinction
record contains a 26 Myr periodicity.' Since they could not
conceive of any terrestrial process that would be so periodic
they were inclined to favour an extraterrestrial cause.

Well before Raup and Sepkoski's paper was published
many people knew about their astonishing findings. Sepkoski
had presented the results at a symposium on extinctions held at
Flagstaff, Arizona, the previous autumn, preprints had been
circulated, and the media had been aroused, with articles
appearing in both the scientific and popular press. Thus it was
that only two months after the Raup and Sepkoski paper
appeared in print, the issue of *Nature* for 19 April 1984 contained five papers that were written in direct response to it.
This phenomenon led the distinguished editor of *Nature*, John
Maddox, to complain in that issue about the circulation of
preprints to a select inner circle, cutting out others who

were not in the network. (Raup subsequently endeavoured to exculpate himself from this venal sin by pointing out that the scientists in question had *requested* preprints, having heard of the periodicity work through articles in the press.)

The five papers fell into three categories. Walter Alvarez and his astrophysicist colleague at Berkeley, Richard Muller, had teamed up to subject the record of impact craters on the Earth to time-series analysis. Only 13 craters, of a much larger number of known craters, were dated adequately enough but these were claimed to be sufficient to enable a 28-million-year periodicity to be detected. Then there were two astronomical hypotheses relating extinction events to comet impact, the galactic plane, and the companion star, each put forward independently by two groups of researchers.

The galactic-plane hypothesis was proposed independently by two groups of American astronomers. It takes account of the fact that the Sun crosses the plane of our galaxy twice in a cycle lasting between about 62 and 67 million years. Close to the galactic plane there is an increased likelihood of encountering giant clouds of gas and dust, known as the molecular clouds. These could perturb the cloud of numerous comets close to the Solar System known as the Oort cloud – widely accepted by astronomers though never observed – causing a few of those comets to collide with the Earth. Several of these comets could perhaps hit the Earth in a period of time up to about a million years, with disastrous consequences for many organisms. Periodic extinctions might thus be explained, although the period would be nearer to 30 than 26 million years.

The alternative hypothesis, by two other groups of American astronomers including Richard Muller, postulated that the

Sun had an unseen companion star, subsequently dubbed 'Nemesis' after the ancient Greek goddess who ensured that no mere mortal ever challenged the dominance of the gods. This star had a highly eccentric orbit. When near the perihelion (the point at which it was nearest to the Sun) it was supposed to perturb the orbits of comets in the Oort cloud, thereby initiating an intense comet shower upon the Earth. The end result from the Earth's point of view was thus much the same, whichever hypothesis might be preferred.

Until this time I had been an interested bystander concerning the extinctions controversy, but I began to be drawn into it by accepting an editorial invitation to contribute a 'News and Views' piece for *Nature*, commenting on the five papers and the article that had started it all. My general reaction was one of scepticism. I questioned the reliability of the geological timescale used by Raup and Sepkoski. Because of the poor quality of many of the data on which it was based, and the large errors involved, the use of other timescales might have found no periodicity. I also noted that some of their extinction events were extremely dubious; others were minor, and deviated from the purported periodicity by several million years. The argument for periodic cratering through the Phanerozoic, based on so few data points, was not readily believable. What I reacted to most strongly was that a group of astronomers seemed to be blithely entering the extinctions debate provoked by the original Alvarez paper without making any attempt to learn more about what geology had to say on the subject; for instance about global changes of sea level and climate. 'Before astronomers indulge in further speculations about the cause of mass extinctions they would do well to learn something about the rich stratigraphic record of their own planet.' I was left with the impression of a science in which too many theoreticians were chasing too few facts, a situation very different from geology.

Two different controversies were generated as a result of Raup and Sepkoski's article and its immediate aftermath. The first of these concerned the rival merits of the astronomical hypotheses. The Nemesis hypothesis was criticized on two counts: first, because the companion star has never been observed (and still has not, despite years of intensive searching by Muller); and secondly, because the orbit of the supposed companion star would be unstable because of the gravitational deflection induced by passing stars; thus there could be no periodicity. The galactic-plane hypothesis is less easy to dismiss, and the well-established periodicity is indeed tantalizingly close to that claimed by Raup and Sepkoski. However, it faces the serious problem that the Sun is currently very close to the galactic plane but the last purported extinction event was 11 million years ago. According to the hypothesis the Sun should now be at the maximum distance from the galactic plane. A group of British astronomers have pointed out another serious if not insuperable difficulty with the hypothesis: the molecular clouds required for cometary perturbation are too sparsely distributed to make encounters with the Oort cloud plausible.

The second controversy concerned the extinction periodicity. An exciting result based on statistics is bound to attract statisticians into the fray, and two University of Chicago colleagues of Raup and Sepkoski duly obliged. Stigler and Wagner observed that a major component of their extinction periodicity analysis was a significance test that decisively rejected the alternative hypothesis that extinctions occurred randomly. Stigler and Wagner confirmed this result, but discussed two things that led them to conclude that the apparent periodicity could be a statistical artefact. Certain types of error in measurement can enhance a periodic signal or cause a pseudoperiodic signal to emerge from data that are aperiodic. The 'hypothesis of a periodic dynamic structure is so powerful in its implications, and so selective in the case with which it imposes itself on us with limited data sets such as this one, that it must be required to pass a stringent test.'

Raul and Sepkoski countered my criticism of their use of a particular timescale by repeating their analysis with other timescales, and showing that the periodicity persisted, although the error margins were greater. Later use of improved timescales, such as the Gradstein timescale for the Mesozoic and Harland's timescale for the Cenozoic, has however considerably weakened Raup and Sepkoski's case. The best fit using these newer and better timescales occurs with a periodicity of 20.5 million years centred on the end-Triassic and early Toarcian extinction events. This produces a 50 per cent success rate; only six of the predicted twelve mass extinctions occur at the predicted times. A fit to a 28.5-million-year periodicity centred on the end-Cretaceous and end-Cenomanian events most closely approximates to the periodicity originally proposed by Raup and Sepkoski, although only four of a predicted nine mass extinctions occur near the correct time. Furthermore, as indicated in my 1999 book with Paul Wignall, there are strong reasons for doubting the reality of a number of Raup and Sepkoski's claimed mass-extinction events in the mid- and end-Jurassic, early Cretaceous, late Eocene, and mid-Miocene. An independent compilation of all Phanerozoic families, including terrestrial ones, under the editorship of

Mike Benton of the University of Bristol enabled Benton to calculate his own extinction metrics. His cogent conclusion is that 'The present data do not lend strong support to this idea [of periodic mass extinctions], because only six, or perhaps seven, of the events are evident.'

Raup and Sepkoski themselves admitted that they could find no evidence in their analysis of Sepkoski's database for any periodicity in the Palaeozoic. Nowadays there is little talk in the palaeobiological community of any periodicity in mass-extinction events. Their occurrence can be considered with confidence as being random, but I prefer the term 'episodic'.

The likeliest causes

One thing that should be apparent by now is that it is extremely unlikely that there is a single overarching cause of mass extinctions, as propounded in the past by various scientists, such as Newell (marine regression), Stanley (climatic cooling), Raup (bolide impact), or Courtillot (volcanism), although all these factors seem to be implicated to some degree. **Figure 9.1** presents in tabular form a list of the greater and lesser events that have been mentioned in this book, with in addition an event at the end of the Precambrian, which is mentioned in the Hallam and Wignall book. The late Eocene event is included, although it has been pointed out that this event, though important and related to climatic cooling, was anything but catastrophic, having extended over several million years in time. Most of the events appear to be associated with more than one factor. The end-Ordovician event is linked to both climatic cooling and warming, and with both marine regression and transgression or anoxia, or both. This is because it is a double event: the earlier one is thought to be related to glacially induced cooling and regression, the later to warming as polar ice disappeared, and to a rise of sea level associated with the spread of anoxic waters.

The commonest association is with volcanism, climatic warming, and anoxia or marine transgression, or both. This is most strikingly the case for the end-Permian mass extinction, the biggest of all. The pronounced negative shift in the carbon isotope curve recognized at this time is most plausibly interpreted as due to the release into sea water of methane from methane hydrate. This raises an important question concerning the comparably large shift at the end of the Cretaceous, which has been generally assumed to be the consequence of a drastic fall in the primary productivity of the oceans after a bolide impact – the 'Strangelove Ocean'. Could a large component of this striking geochemical signal be due instead to the breakdown of methane hydrate? Perhaps one of the main reasons why the end-Permian marked the most significant of all mass extinctions is that a 'runaway greenhouse' was created before a processes of negative feedback restored normality. Volcanism triggered the release of methane, and this in turn contributed through the greenhouse effect to global warming, adding to the effect already produced by volcanic emissions of carbon dioxide into the atmosphere. Increased anoxia would also be provoked by a rise in temperature, owing to the fact that the solubility of oxygen is lower in warmer water.

What is most evidently apparent from Fig. 9.1 is the general insignificance of bolide impact, apart from the one event at the end of the Cretaceous. Even here, impact was not the whole

	Bolide impact	Volcanism	Cooling	Warming	Regression	Anoxia/ transgression
Late Precambrian						●
Late Early Cambrian					●	●
Late Cambrian biomeres			○			●
End-Ordovician			●	●	●	●
Frasnian-Famennian			○		○	●
Devonian–Carboniferous			○			●
Late Guadaloupian					●	
End-Permian		●		●	○	●
End-Triassic		●		○	●	○
Early Toarcian		●		○		●
Cenomanian–Turonian			○			●
End-Cretaceous	●	●	●		●	○
End-Palaeocene		●		●		●
Late Eocene			●			

● Probable link ○ Possible link

Fig. 9.1 Summary of the proposed causes of the main Phanerozoic mass-extinction events.

story of the mass extinction at this time but merely the culminating *coup de grâce*, albeit a spectacularly catastrophic and important event. The Earth has undoubtedly endured in Phanerozoic time a succession of impacts from outer space, as the cratering record shows, but they seem to have disturbed the biota on a global scale remarkably little. If this book has a fundamental message it is that biotic catastrophes and calamities have their origins for the most part in entirely Earthbound causes, which tie up with events in the mantle.

This does not mean that we yet have all the answers; far from it. The great Maurice Ewing, founding father of one of the world's greatest institutes of geological oceanography, the Lamont-Doherty Geological Observatory of Columbia University, New York, used to say in the pioneering days of his subject, 'We know more about the backside of the moon than we do about the bottom of the sea.' We now know much more about the bottom of the ocean, but we remain in deep ignorance about the mantle and its relationship to the core.

Most of our information must necessarily come indirectly, from three areas of research: from seismology, and especially from the branch of seismology called seismic tomography; from research on the petrology and geochemistry of igneous rocks at the Earth's surface or at shallow depths within the crust; and even more indirectly from studies of meteorites. Mention was made in Chapter 8 of the likely relationship between the eruption of flood basalts at the surface and plumes rising through the Earth's mantle from near the core to the base of the lithosphere. Such plumes are poorly understood, and are controversial in some circles, but they appear to be the best model we have at present. There is a strong likelihood that the rise of plumes through the mantle, related to zones of increased temperature, is episodic, with intervals of a few tens of millions of years. The effects at the Earth's surface can be various. Volcanism, and its possible climatic implications, is only the most obvious. Large sectors of the Earth's crust may be subjected to upheavals and subsequent collapse after the eruption of lavas. Such *epeirogenic movements*, as geologists call them, can have significant effects, at least on a regional scale, on sea level. Moreover, major plate-tectonic movement involving lateral displacement may be initiated. Judging from the stratigraphic record, it seems as though from time to time all hell breaks loose after long periods of quiescence. These episodes I like to call *geohiccups*.

The marked rise of sea level close to the beginning of the Triassic period poses a particularly interesting problem. There was no significant plate-tectonic activity at this time, and in any case the global sea-level changes caused by plate tectonics seem to be much too slow for what in geological terms was an exceptionally rapid rise. Nor can a rise due to the melting of polar ice caps plausibly be invoked, because such ice caps had evidently disappeared well before the end of the Permian. A possible explanation emerges by comparison with events within the mid-Cretaceous, with the rise of what has been called a *superplume* beneath the Pacific Ocean. This superplume was the prime agent responsible for one of the biggest global marine transgressions on record, as the rising submarine domes or plateaux of basaltic provinces displaced sea water over the continents. The clear implication of this comparison is that, apart from the terrestrial eruptions of the Siberian Traps, there were also substantial submarine basaltic eruptions at the time of the Permian–Triassic boundary. These would contribute significantly to the greenhouse effect at that time as carbon dioxide was erupted into sea water, eventually finding its way into the atmosphere (carbon dioxide, like oxygen, has a lower solubility in warmer water). The evidential support for the Cretaceous story is good because we have a detailed record of Cretaceous rocks under the present ocean floor. Unfortunately there is no such record for the Permian, because oceanic rocks of that age have either been subsequently subducted beneath the continents or accreted to their margins by plate-tectonic processes.

Plate tectonics is thought by many to be the prime control on the coming and going of major ice ages in geological history. For example, a polar ice cap cannot develop unless a major continent is situated over one or both poles. Plate movements cannot, however, provide the whole answer because there was no significant shift of continental positions during the Permian, when the Gondwana ice cap disappeared. In more recent times, the dramatic global climatic cooling that took place in the late Cenozoic, culminating in the Quaternary ice age, has been related primarily to the uplift of the Tibetan Plateau and the Himalayas after the collision of India, originally part of Gondwana, with Asia. The uplift at this time of such areas of high altitude, together with others such as the American West and the Andes, is thought to have significantly increased the amount of continental weathering, and this has an important climatic consequence for the world as a whole.

Increased weathering implies an increased extraction of carbon dioxide from the atmosphere, as carbonated water attacks the rocks and the dissolved calcium and bicarbonate ions eventually react to produce limestone. Thus a reversed greenhouse effect results. Like much else in geology, this is a good idea that as yet lacks conclusive support, but it is probably the best model available at present. Such an interpretation could also be relevant to the good correlation that exists between the eruption of large flood-basalt provinces and mass extinction at a time when the present continents were part of a coherent supercontinent, Pangaea. This continent does not appear to have been subjected to significant uplift on an extensive regional scale, and its climatic regime was largely arid. Under such circumstances, removal (draw-down) of carbon dioxide from the atmosphere by weathering processes would have been minimal, with the result that carbon dioxide derived from volcanic eruptions would have remained in the atmosphere.

Apart from possible ultimate driving mechanisms, about which we still have much to learn, there remains the problem of what environmental changes can be so drastic on a global scale as to make a large proportion of the world's biota extinct in a geologically brief time. A major reason why Raup has been inclined to support bolide impact as the most probable cause is that he could not conceive of any merely Earth-bound cause that would do the trick. Against this it has been argued in this book that the empirical evidence of close temporal association does indeed support a terrestrial rather than extraterrestrial cause; but this is not, of course, an adequate explanation. We still lack reliable quantitative information about the rate and amount of environmental change, of whichever sort. More needs to be known, not just about particular organic groups, but about whole ecosystems, before we can aspire to a fuller understanding. One important point at least should be stressed here. A change of environment of moderate magnitude may be more important for mass extinction than one of extreme magnitude provided that the change is sustained long enough. This could be a prime reason why, the baleful influence of our own species apart, the Quaternary has experienced low extinction rates despite the strong fluctuations of temperature.

To conclude, much remains poorly understood about the causes of mass extinctions, but rather than bemoan this fact I would prefer to salute the significant progress in understanding that has been achieved in recent years, utilizing techniques from a variety of research fields. To me the glass is half full rather than half empty.

10

The evolutionary significance of mass extinctions

Darwin was firmly of the opinion that biotic interactions, such as competition for food and space – the 'struggle for existence' – were of considerably greater importance in promoting evolution and extinction than changes in the physical environment. This is clearly brought out by this quotation from *The Origin of Species*:

> Species are produced and exterminated by slowly acting causes . . . and the most important of all causes of organic change is one that is almost independent of altered . . . physical conditions, namely the mutual relation of organism to organism – the improvement of one organism entailing the improvement or extermination of others.

The driving force of competition in a crowded world is also stressed in another quotation presenting Darwin's famous wedge metaphor:

> In looking at Nature, it is most necessary . . . never to forget that every single organic being around us may be said to be striving to the utmost to increase in numbers; that each lives by a

struggle at some period of its life; that heavy destruction inevitably falls either on the young or the old, during each generation . . . The face of Nature may be compared to a yielding surface, with ten thousand sharp wedges packed close together and driven inwards by incessant blows, sometimes one wedge being struck, and then another with greater force.

The implication of the Darwinian view concerning the dominance of biotic competition is that for each winner there is a loser – a king of zero-sum games. It has been accepted more or less uncritically by generations of evolutionary biologists, but not until the 1970s did it become graced with a name – the *Red Queen hypothesis*. The story behind the emergence of this name is an interesting one.

At the beginning of the 1970s the rather eccentric University of Chicago palaeobiologist Leigh Van Valen did some interesting research concerning the analysis of survivors of Phanerozoic taxa which suggested that the probability of a fossil group becoming extinct was more or less constant in time. To account for this, Van Valen put forward his Red Queen hypothesis. Readers of Lewis Carroll's *Through the Looking-Glass* will recall that the Red Queen explained to Alice that where *she* lived it took all the running one could do to stay in the same place. Van Valen's hypothesis is based on the assumption that all species within a given adaptive zone compete intensively. A successful adaptive response by one species is assumed to occur at the expense of other species, which must, as the 'quality' of their environment is reduced, either themselves adapt by speciating (i.e. evolving into new species) or becoming extinct. This phenomenon leads to an endless chain of adaptive responses, and in the long run means that fitness and rate of extinction remain constant.

Not unnaturally, Van Valen thought that his work was worthy of the pages of either *Science* or *Nature*. Unfortunately, the paper was rejected by both these illustrious journals. This could have discouraged many, but Van Valen simply went ahead to create his own journal, *Evolutionary Theory*, and published his paper in the first issue. Although his data and conclusions have been disputed, the term 'Red Queen' has had an astonishing success and is now firmly established in the evolutionary literature. (Incidentally, Van Valen also published another, satirical, journal on an occasional basis, called the *Journal of Insignificant Research*. My preferred part was that devoted to papers in the scientific literature with curious titles and authors' names. Two of my favourites were (1) 'The fertility of witches in the Firth of Clyde', published by the very respectable *Journal of the Marine Biological Association*, and (2) 'The zoological perspective in social science', by Lionel Tiger and Robin Fox.)

Van Valen has a specialist knowledge of fossil mammals. The high rate of diversification and evolutionary turnover in mammals, which makes them, like ammonites, valuable biostratigraphic indicators, is likely to be the result of a variety of factors, such as strong competitive interactions leading to specialization in feeding methods, limitations on food supply, high mobility and use of energy, interspecies aggression, and territoriality. Such factors will conspire to lower the 'resource threshold' that is needed to prevent extinction, as compared with other animals. Although the Red Queen model might well apply to mammals, there are doubts about its more general validity. Thus, in sharp contrast to mammals, the bivalves, that is, clams and oysters, are nearly all sea-bed suspension feeders which mind their own business, are characterized by weak interactions with other species, exhibit primitive inflexible behaviour, have an uncrowded, largely sedentary, mode of life, and have generalized feeding habits. As a direct consequence of this pattern they have substantially lower rates of evolution than mammals. Limits on bivalve populations are imposed more by predation and fluctuations in the physical environment than by food resources, and biological competition is minimal. What is true for bivalves is without much doubt true of the majority of benthic invertebrates, which make up a large part of the fossil record. If we are to cite Lewis Carroll, *Through the Looking-Glass* might be less appropriate as a textual source for the relevant evolutionary sermon than *The Walrus and the Carpenter*.

The alternative to the Red Queen model, stimulated by the study of mass extinctions, has been termed the *Stationary model* by the doyen of British evolutionary biology, John Maynard Smith, and his Norwegian colleague Nils Stenseth. This implies that, contrary to Darwin's belief, the prime motor of evolutionary change is the physical, not the biotic, environment. Indeed, evolution could conceivably grind to a halt in the absence of abiotic change. Testing these alternative models from the fossil record has not proved straightforward. The most promising material for this purpose is a variety of Cenozoic microfossils from deep-sea cores, such as foraminifera, radiolaria, and diatoms. The results unfortunately have proved rather inconclusive so far. One group of workers thought that their evidence marginally favoured the Red Queen model, but others countered this by pointing out evidence of physical oceanographic change that had been ignored.

Since biotic competition manifestly exists today we need to enquire further what sort of competition may have predominated in the past, in the light of what is now understood of biotic turnover through longer and shorter periods of time.

The role of competition as perceived from the stratigraphic record

The type of competition that Darwin had in mind might be termed *displacive competition*, implying dynamic behaviour on the part of the newly arrived species or higher taxon in the wedge. How could this be recognized in the stratigraphic record? One reasonable inference is that after the new taxon first appeared, it might progressively expand its abundance and diversity concomitant with the progressive reduction of its biologically most closely related or potentially competitive rival, known as the *double wedge pattern* (**Fig. 10.1a**). An alternative type of competition is called *pre-emptive competition*, which implies that evolutionary success favours the incumbents of a particular ecological niche. The earlier species occupant of a niche would remain there until some physical disturbance caused its elimination. Only then would other biologically or ecologically related forms opportunistically occupy the niche. As portrayed in **Fig. 10.1b**, there would be very little overlap in time of the earlier and later taxa as portrayed in the stratigraphic record.

What does the record actually show? Let us take the familiar case of dinosaurs and mammals, which both emerged at about the same time in the late Triassic, about 230 million years ago. There is no indication that the 'biologically superior' mammals progressively out-competed the dinosaurs throughout their

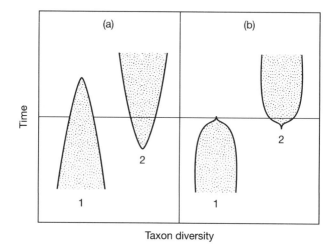

Fig. 10.1a+b What the fossil record in a stratal succession may reveal for (a) displacive and (b) pre-emptive competition.

long history together. Rather it would appear that the two groups coexisted in different ecological niches, the mammals remaining low in diversity and small in size – no larger than rats – but with nocturnal vision, both characteristics making possible the exploitation of niches not in direct competiton with the dinosaurs. Very soon after the dinosaurs were finally killed off by the end-Cretaceous mass-extinction event, however, the mammals greatly expanded their diversity, individual size, and occupation of a variety of ecological niches, some of which, such as the air and the sea, had never been occupied by dinosaurs. Within a few million years of the K–T boundary all the principal mammalian orders had become established, and in the course of time (in the Miocene) huge beasts rivalling some of the larger dinosaurs in size had evolved. This pattern of change seems to be a spectacular example favouring pre-emptive competition.

Triassic reptiles provide another likely example that has been put forward by Mike Benton. The conventional view among experts had been that the reptiles that earlier evolved had been progressively out-competed by the dinosaurs. This view is not supported by the stratigraphic record, which shows that the older reptiles had become extinct before the earliest dinosaurs were on the scene. All this took place before the end-Triassic mass extinction, but it is associated with at least one extinction event of lesser magnitude. What is true for land vertebrates appears to be more generally true for the marine invertebrate record. To take just the example of the group with the highest rate of evolutionary turnover in time, the ammonoids (including the ammonites), a series of families in the late Palaeozoic and Mesozoic, both radiated and became extinct in a more or less regular succession, with little or no overlap in time. This phenomenon has been termed an *evolutionary relay*.

It has been claimed that the turnover of plant taxa with time exhibits a fundamentally different pattern from that of animals because plants are less vulnerable to mass-extinction episodes, and displacive competition plays a major role. However, Bill DiMichele of the Smithsonian Institution in Washington D.C. disputes this for the Carboniferous to Permian vegetational transition, which he considers to be replacive rather than displacive. Taxa that originated in peripheral, drier habitats in the late Palaeozoic tend to be subgroups of seed plants. The life histories of seed plants indicate *a priori* a greater resistance to extinction than most groups of 'lower' vascular plants, such as ferns, lycopsids, and sphenopsids. This prediction is confirmed by the nearly continuous expansion of seed-plant diversity since the Palaeozoic, at the expense of lower vascular plants.

Thus the culling by both major and minor extinction events of organisms that evolved earlier throughout Phanerozoic time could have provided the opportunity for organisms that evolved later to take over their ecological niches, a pattern more in favour of the Stationary than the Red Queen model. Stephen Jay Gould has made great play of how much the course of evolution has been interrupted by such contingent historical events as mass extinctions, the bigger events perhaps being so significant as to 'reset the evolutionary clock'. This forms part of Gould's challenge to the conventional view of evolution as a progressive one leading eventually to the emergence of our own species. This is a very large topic that goes beyond the scope of this book, but one or two points are worth making briefly here.

Evolutionary success can be measured in various ways, and by some criteria bacteria, traditionally regarded – rather inaccurately – by us as exceptionally simple organisms, are the most successful in terms of biomass, the variety of metabolisms utilized, and the range of environments occupied. No obvious progress through time is either discernible or likely. So far as the complexity of neural systems culminating in the human brain is concerned there has, however, undoubtedly been progress through time. Simon Conway Morris of the University of Cambridge goes further and points to the ubiquity of evolutionary convergence as providing a strong challenge to Gould's views. The example of the convergent, streamlined shape of active marine predators in different groups of vertebrates is a familiar example, such as tuna and sharks (fish), ichthyosaurs (reptiles), and dolphins (mammals), but convergences of all sorts, including those at the molecular level, can be recognized throughout the organic world. They include progressive tendencies in a variety of evolutionary lineages towards higher intelligence, traditionally regarded as the epitome of evolutionary success.

Evolutionary biologists might be troubled by the general rejection of displacive in favour of pre-emptive evolution as the general pattern to be discerned from the fossil record as a whole. What, they might ask, about co-evolution? The co-evolution of plants and insects could well be cited as a superb example. One thinks in particular of the marvellous adaptations of orchids for particular insect pollinators, the study of which was pioneered by none other than Charles Darwin. Surely such co-evolution began to evolve as early as the Cretaceous, when the angiosperms became the dominant land flora? No doubt it did, but there is unsurprisingly no fossil evidence for it. There is one example from the fossil record, though, that appears to be an excellent example of co-evolution.

Geerat Vermeij, of the University of California at Davis, known to everyone as Gary, is a remarkable man. He has written several books and many papers, all of which display an immensely wide range of knowledge handled with a high level of intellectual sophistication. However, uniquely for a palaeobiologist, none of his writings contains either pictures or diagrams. This is because he is blind. He has a highly retentive memory and depends on others, mainly his wife, to read relevant scientific literature to him. Not only is he an expert on the taxonomy of snails, his favourite creatures, but he also knows a great deal about grasses, although he has to depend in both cases on his tactile sense. How many of us, even with our visual sense intact, can tell different types of grasses apart?

Vermeij is best known for his theory of evolutionary escalation. Concomitant with the emergence and radiation in the Cretaceous of sophisticated benthic predators, such as teleost fish, carnivorous neogastropods, and crabs, there was a change in the shells of two groups that were part of their favoured molluscan prey: the gastropods and bivalves. Some developed thicker shells or acquired protective spines or strong ribs. (Palaeontologists are apt to speak informally of such features as 'ornament', but there was nothing frivolous in this for the molluscs; it was rather a matter of life and death.) Others, among the bivalves, developed a greater capacity to burrow deeply in the sediment out of harm's way. Yet others retreated to the deep sea. The predators evolved in turn to become more efficient in capturing their prey, something which appears to be equivalent to what Richard Dawkins has called an 'evolutionary arms race'. One thinks in the modern world of the extraordinarily sophisticated use by plants of poisons and spines to resist animal feeding and the corresponding sophisticated response on the part of the animals.

There is no reason, therefore, to deny conventional Darwinian evolution throughout geological history to account for the numerous fine adaptations of organisms, but the more general patterns of change seem more likely on present evidence to be controlled by contingencies relating to perturbations of the physical environment, the most important of which have led to mass extinctions.

The derivation of evolutionary patterns from mass originations and mass extinctions

The significance of mass extinctions within the broader pattern shown by the history of life can be appreciated only in the context of diversity changes, which take into account originations as well as extinctions. 'Diversity' has various meanings in biology, but the simplest one that is applicable in geological history is a straightforward taxon count. Until recently the only database available was Sepkoski's compilation of Phanerozoic marine families and genera, but with the second volume of *The Fossil Record* under the editorship of Mike Benton we have a full record of terrestrial as well as marine families. Benton has been active, with mathematically inclined biologists, in exploring the patterns that can be discerned for both groups.

Diversification on the broader scale, in which new groups of organisms arise and become extinct, is referred to by biologists as 'macrodiversification'. Of the various models of macro-diversification patterns, those based on equilibria seem to be more easily accepted because of the apparently widespread applicability of systems studied by ecologists, where the time-scales are exceptionally short by geological standards. Their applicability to changes over geological timescales is more controversial. The classic equilibrium model is represented by a logistic curve which can be produced by a diversity-dependent process leading to the formation of a plateau (**Fig. 10.2a**), which is interpreted as representing the global carrying capacity of the environment. The widely accepted notion of biological and evolutionary equilibria is based on the belief that there is a fixed number of ecological niches and that life expands to fill the available slots. The equilibrium model assumes that after the initial filling of ecospace, newly evolved taxa with superior adaptations always replace earlier taxa, or drive them to extinction, so that a general and constant total species diversity is always maintained. In other words, the 'carrying capacity' of the environment is fixed, and diversity will increase after an original equilibrium is established only after life has invaded major new sets of habitats. These include the invasion of muddy sea floors in the Ordovician, the conquest of the land by plants and arthropods in the Silurian and Devonian, and the radiation of the angiosperms in the Cretaceous, with its concomitant effect upon insects.

The alternative model is represented by an exponential curve (**Fig. 10.2b**), which is derived most simply from constant doubling of numbers, with no hint of a plateau (a curve that accelerates at a rate less than doubling is called a damped exponential). Exponential curves are ubiquitous in studies of the reproduction of individual populations of organisms

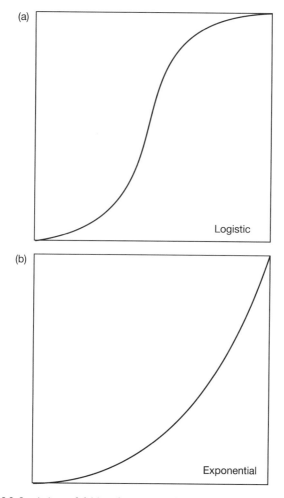

Fig. 10.2 Logistic model (a) and exponential model (b) to be tested against data from the fossil record.

(think rabbits!) and no plateau is possible. If populations do not continue to multiply at an exponential rate it is because natural checks are applied, such as starvation and predation. The exponential model carries the implication that there is no limit to increase in diversity, which will continue to grow at an ever-increasing rate; in other words, the notion of carrying capacity does not apply. Biological innovations produced in the course of evolution will in effect create new ecological niches. Examples include, among plants, the development of an arborescent (treelike) habit in the Devonian and Carboniferous. Among the animals a variety of innovations readily present themselves in, for example, the piercing and crunching of shells by marine predators, and flight and tree-climbing among terrestrial inhabitants.

These models must of course be tested against the fossil record. **Figure 10.3** presents the diversity change of marine and terrestrial families through time, from the latest Precambrian to the present. There is of course no record of land animals or plants prior to the Silurian, but the record of marine organisms can be carried back to the peculiar Vendian fauna at the end of the Precambrian. Sepkoski did not consider this Vendian fauna in his analyses, but he considered that the diversity changes through the Palaeozoic are consistent with an equilibrium model, with a sharp rise in the Cambrian and early Ordovician subsequently flattening out to a plateau. This pattern does indeed seem evident for both orders and families, but is less clear from Sepkoski's unpublished compilation of genera, which suggests that it may break down at the level of species. There are unfortunately no reliable data on species. Subsequently, however, in the Mesozoic and Cenozoic, in both the Sepkoski and Benton compilations the picture is one of an

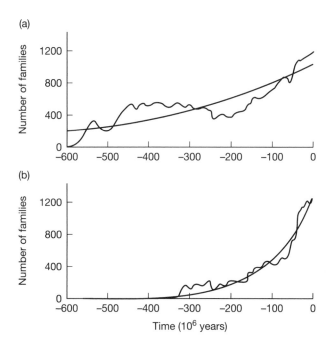

Fig. 10.3 The changing diversity of (a) marine and (b) continental
families through the Phanerozoic and back into the latest
Precambrian (Vendian). In both (a) and (b) an exponential
curve extending back to 600 million years is portrayed. For
continental organisms this curve is closely coincident with
one for the past 250 million years. (Simplified from Hewzulla
et al. (1999).)

ever more rapid rate of increase in diversity, which is more consistent with an exponential model. This result is challenged, however, by Mike Foote of the University of Chicago, whose comprehensive statistical analysis of Sepkoski's generic database suggests that his results support logistic rather than exponential modelling of biological diversification. The terrestrial record strikingly matches an exponential model, as shown in Fig. 10.3.

Mass originations and mass extinctions show up respectively as positive and negative deviations from the curve portrayed in Fig. 10.3, which is an exponential fit for the data from 600 million years ago onwards. The largest mass origination relates to the Cambrian 'explosion' of multicellular animals, which remains one of the most striking and poorly understood phenomena in the history of life. The end-Permian mass extinction is responsible for a large negative blip in both the terrestrial and marine data. On the other hand, the end-Cretaceous mass extinction, although clearly recognizable in both the marine and terrestrial data, seems to represent no more than a minor blip in a more general pattern of increasing diversity. The remainder of the 'big five' show up clearly only in more detailed presentations of the data.

The general patterns portrayed do not provide strong support for Gould's view that mass extinctions have substantially altered the course of history, or have 'reset the ecological clock'. Consider just two of the changes that took place as a result of the biggest mass-extinction event. The rugose corals had first arisen in the early Palaeozoic, and were a dominant reef-building group until their extinction at the end of the Permian. Reef ecosystems appear to have disappeared for several million years after the end of the Permian, but in the mid-Triassic a new reef-building coral group, the Scleractinia, appeared. Structurally this group is quite distinct from the Rugosa, and it is generally thought that the scleractinians evolved from a related, softer-bodied group of cnidarians, to fill the niche left empty by the demise of the Rugosa. The second change concerns bivalved organisms that dominated the shelly macrobenthos at various times. The brachiopods were dominant throughout most of the Palaeozoic in terms of both abundance and diversity, but after the complete extinction or near-extinction of several major groups at the end of the Permian they lost their importance. Some still survive today but as very subordinate and geographically restricted components of the benthos. The bivalve molluscs were present as subordinate components of the benthos from early Ordovician times onwards but did not significantly expand their abundance and diversity until the mid-Triassic onwards. Throughout the Cenozoic, until today, they have dominated, together with gastropods, the shelly macrobenthos. Both groups are almost exclusively passive, suspension-feeding organisms living on the sea bottom, and it could be argued that the bivalves have out-competed the brachiopods by a Darwinian displacive competition. However, as Gould pointed out in a paper some years ago, the bivalves did not 'take off' until after the brachiopods had been so heavily hit by the end-Permian extinctions. The implication is that any competition between the two groups was preemptive rather than displacive. As Gould stated in his paper, brachiopods and bivalves were probably, in Longfellow's phrase, merely 'ships that pass in the night'. Bivalves survived the end-Permian extinction event better, and refilled the niche substantially vacated by the brachiopods. Such events as these are very interesting for the palaeobiologist, but cannot be said to be of fundamental significance in understanding the course of evolutionary history.

This course, as it can be discerned from the fossil record, is essentially as follows. Animals with skeletons first appeared in the Cambrian (archaeocyathan sponges, inarticulate brachiopods, trilobites, vertebrates), followed by corals, articulate brachiopods, cephalopods, ostracods, crinoids, starfish, and graptolites in the rest of the Palaeozoic. Bivalves, gastropods, malacostracan crustaceans, echinoids, bony fish and marine reptiles became important in the Mesozoic. Except for the reptiles, these marine groups continued to diversify in the Cenozoic. On land, vascularized plants appeared together with arthropods in the Silurian and Devonian; gymnosperms appeared in the Carboniferous, angiosperms in the Cretaceous. The first limbed animals arose in the Palaeozoic; insects in the late Palaeozoic, and social insects in the Cretaceous. Among the vertebrates, amphibians appeared in the Devonian, reptiles in the Carboniferous, and pterosaurs and dinosaurs in the Triassic and Jurassic. Birds first arose in the late Jurassic and mammals became dominant in the Cenozoic. Only two of these events appear to be directly related to mass-extinction events: the change of marine invertebrate faunas from the Palaeozoic to the Mesozoic (for example, the change from brachiopod to bivalve dominance), and the replacement of dinosaurs by mammals as dominant land vertebrates. Using his *Fossil Record* database, Benton estimates that the average family extinction rate at the end of the Permian was 61 per cent, 63 per cent for continental, and 49 per cent for marine organisms. The family extinction rates for the marine record as a whole are close to those of Sepkoski.

The most interesting result of Foote's statistical analysis of Sepkoski's data is that changes in generic diversity within higher taxa of marine animals on the temporal scale of a few million years are more strongly correlated with changes in extinction rate than changes in origination rate during the Palaeozoic. After the Palaeozoic the relative roles of origination and extinction in diversity dynamics are reversed. Metazoa as well as individual higher taxa shift from one mode of diversity dynamics to the other. This phenomenon clearly requires an explanation. Assuming that physical perturbations did not sharply decrease in magnitude or frequency from the Palaeozoic to the Mesozoic, the observed difference in diversity dynamics seems to suggest a change in the way that ecosystem structure allowed shocks to percolate through the biota. Such change could involve increasing niche specialization and greater energy flow. How these and other aspects of ecosystem evolution have contributed to the evolution of diversity dynamics is yet to be determined.

The most striking increase in diversity took place in the Cenozoic. The steepest latitudinal and longitudinal gradients in taxonomic diversity at the present day are those associated with tropical foci of high diversity. Within the marine realm, a uniform, pan-tropical fauna was progressively disrupted by a series of plate-tectonic events, the most important of which were the early Miocene collision of Africa/Arabia with Europe and Australia/New Guinea with Indonesia, and the Middle Miocene–latest Pliocene rise of the Central America Isthmus between North and South America, which had the effect of separating the Caribbean from the East Pacific. This had the net effect of establishing two main tropical foci of high diversity: the Indo-West Pacific and the Atlantic/Caribbean–East

Pacific. Together with the physical isolation of Antarctica, these same tectonic events contributed significantly to global cooling throughout the Cenozoic era. This in turn led to the imposition of a series of thermally defined provinces, and thus to a considerable degree of biotic differentiation, involving speciation, on a regional scale.

Finally, we should take due note of the recently discovered phenomenon of *self-organized criticality*. The theory was originally developed from the study of avalanching piles of sand. One specific trigger, which at the limit can be merely a single grain added to the pile, can initiate avalanches of a wide range of magnitudes. The magnitude of the avalanches obeys a power law; that is, there are a few large avalanches and many small ones. The relative magnitude of mass extinctions also appears to approximate to such a power law, as do other natural phenomena such as earthquakes and floods. One key implication is that there is not necessarily a simple relationship between the size of a given mass extinction and the size of the causal factor. Furthermore, there may sometimes have been a combination of interacting factors that exceeded some critical threshold, possibly with cascading effects. Self-organized criticality signifies intrinsic non-linear dynamics. Mathematically this implies fractal statistics and scale-free dynamics. The results obtained so far are consistent with the idea that the 'big five' are at the skewed end of a continuous distribution of extinction events of varying intensity. A picture emerges in which many species are evolving or interacting, and this seems to be more consistent with processes of macroevolution than deterministic approaches based on chaos dynamics.

The claim has, indeed, been made on the basis of these mathematical results that biotic interactions play a prominent role in mass extinctions. This is very likely to be true in some instances – indeed it would be hard to deny for integrated ecosystems – but there are obvious limits to such claims if they are intended to provide a general answer. For example, it is impossible to conceive that there was any biological relationship between planktonic foraminifera and dinosaurs that caused them both to suffer mass extinction at the end of the Cretaceous. On the other hand, the extinction of plankton, at the base of the marine food chain, is very likely to be implicated in the extinction of benthos, but this does not qualify as a biotic interaction. A leading theme of this book is that ultimately it is physical changes in the environment that cause mass extinctions, and these changes are not caused by organisms. There is only one exception, our own species, and it is to this that we now turn.

The influence of humans

We saw in Chapters 5 and 7 that the Quaternary was a time of low extinction rates despite a succession of strong environmental changes induced ultimately by climate. This began to change from a few tens of thousands of years ago with the arrival on our planet of *Homo sapiens sapiens*, which can be translated from the Latin as the rather smug 'ultrawise Man'. It is widely accepted today that the Earth is undergoing a loss of species on a scale that would certainly rank in geological terms as a catastrophe, and has indeed, been dubbed 'the sixth mass extinction'. Although the disturbance to the biosphere being created in modern times is more or less entirely attributable to human activity, we must use the best information available from historical, archaeological, and geological records to attempt to determine just when it began.

Extinctions on the land

Towards the end of the last ice age, known in Europe as the Würm and in North America as the Wisconsin, the continents were much richer in large mammals than today: for example, there were mammoths, mastodonts, and giant ground sloths in the Americas; woolly mammoths, elephants, rhinos, giant deer, bison, and hippos in northern Eurasia; and giant marsupials in Australia. Outside Africa most genera of large mammals, defined as exceeding 44 kilograms adult weight, disappeared within the past 100,000 years, an increasing number becoming extinct towards the end of that period. This indicates that there was a significant extinction event near the end of the Pleistocene. This event was not simultaneous across the world, however: it took place later in the Americas than Australia, and Africa and Asia have suffered fewer extinctions than other continents.

There are three reasons for citing humans as the main reason for the late Pleistocene extinctions. First, the extinctions follow the appearance of humans in various parts of the world. Very few of the megafaunal extinctions that took place in the late Pleistocene can definitely be shown to pre-date the arrival of humans. There has, on the other hand, been a sequence of extinctions following human dispersal, culminating most recently on oceanic islands. Second, it was generally only large mammals that became extinct. It is obvious that large animals would make better targets for hunting. They would also have been more vulnerable to the destruction of their habitats induced by humans, because of the combination of their smaller population sizes and lower reproductive rates. Finally, the only reasonable alternative, for the North American extinctions at least, is to invoke environmental changes connected with climate in some way, either directly or indirectly. This seems implausible. The limiting intensity of most of the relevant selective forces, such as cooler air temperatures, simplicity of habitats, and low primary productivity, began to *relax* 18,000 years ago, and the associated environmental changes should not have produced the cluster of extinctions that occurred. The advent of human hunting is the only new factor. The climatic-change hypothesis fails to account for the increasing likelihood of extinctions with increasing body size, the greater severity of extinctions in both North and South America than in Eurasia, the lack of simultaneous extinction in Africa and tropical Asia, and the absence of extinctions at the end of previous Pleistocene glacial periods.

The human overkill hypothesis has nevertheless been criticized on several fronts. The most important criticism is to question why there were no significant late Pleistocene extinctions in Africa, which has a record of hominids going back several million years. This objection has been countered by Paul Martin of the University of Arizona, the principal advocate of the human overkill hypothesis, who points out that there were in fact an appreciable number of extinctions in the early Pleistocene, but few elsewhere in the world, followed by a long period of stability. He speculates that several facts may reflect a co-evolutionary history of humans and other mammals: the highly diverse nature of the surviving African mammals and their adaptation for running; the Pleistocene increase in diversity of antelopes, and the lack of slow, ponderous herbivores like ground sloths and some others in South America and Australia, which became extinct. Some critics have maintained that the survival in Africa of large mammals does not favour Martin's overkill hypothesis, but he uses this fact as an argument against a combination of climatic and cultural causes of extinction.

Another potential problem is that the human predation hypothesis fails to explain the simultaneous extinctions of a number of mammalian or bird species that were not obviously vulnerable to human overkill. To account for this a 'keystone' herbivore hypothesis has been proposed by Norman Owen-Smith of the University of the Witwatersrand, Johannesburg. In present-day Africa elephants, rhinos, and hippos can transform tall grasslands into 'lawns' of more nutritious grasses. The elimination of megaherbivores elsewhere in the world by human hunters at the end of the Pleistocene would have promoted reverse changes in vegetation. These could have been detrimental to the distribution and abundance of smaller herbivores that were dependent upon the nutrient-rich and spatially diverse vegetation created by the impact of megaherbivores.

A fascinating study has been undertaken by John Alroy of the University of California at Santa Barbara. His computer simulation of North American end-Pleistocene human and large herbivore population dynamics, published in 2001, correctly predicts the extinction or survival of 32 out of 41 human prey species. The first solid evidence of large human populations in America – the Clovis culture – is dated at 13,400 years BP (before the present). The key question is whether human population growth could have been sufficiently rapid, and hunting rates sufficiently high, to have driven 73 per cent of large herbivores to extinction. Alroy simulates human population growth and hunting patterns and the population dynamics of 41 large mammalian herbivores across the Pleistocene–Holocene transition. The results of the simulation are unambiguous. Human population growth and hunting almost invariably leads to a major mass extinction. A single best-fit scenario simultaneously makes accurate predictions about extinction outcomes, timing, and human ecology. Furthermore, predicting survival solely on the basis of body mass by declaring all species greater in weight than 180 kilo-grams to be extinct would identify 23 of 37 extinctions and 7 of 11 surviving species. This is significantly better than a random guess.

Times of extinction are another accurate prediction of the best-fit model. The median extinction time is 1229 years after the initial human invasion. It takes 260 years for human popula-tions to exceed 1000 individuals and 410 years to exceed 10,000. So we might not expect the archaeological record to show evidence of humans before this time. Thus a 1000–1200-year overlap of humans and megafauna might be expected. The earliest appearance of Clovis artefacts in the United States is around 13,400 years BP, whereas the youngest radiocarbon dates on the extinct megafauna are around 12,260 years BP. The known overlap of about 1200 years is roughly as predicted. Of a number of parameters that are varied in the course of the simulations, hunting ability is the most important, and mass extinction becomes inevitable given general facts about human ecology. This ecologically realistic model challenges the common-sense notion that no amount of overkill could give rise to a true megafaunal extinction. The overkill model serves as a parable of resource exploitation, providing a clear mech-anism for a geologically instantaneous ecological catastrophe too gradual to be perceived by the people who unleashed it.

The latest work on the Australian megafaunal extinctions also strongly favours a human explanation. All Australian land mammals, reptiles, and birds more than 100 kilograms in weight, and 6 out of 7 genera weighing between 45 and 100 kilograms, died out in the late Pleistocene; 23 out of 24 animals exceeding 45 kilograms became extinct. Radiometric age determinations from nearly thirty fossil sites across the subcontinent indicate an extinction around 46,400 years ago, several thousand years after the presumed first arrival of humans. Climate change, in the form of extreme aridity at the Last Glacial Maximum, might provide a possible explanation. This can, however, be ruled out as a cause of the extinctions, leaving the human alternative, either an overhunting 'blitzkrieg' or an extended period of anthropogenic disruption of the ecosystem. Alroy's model suggests that human overkill might be a sufficient explanation.

While doubts may linger concerning whether a human agency provides a *complete* explanation for the late Pleistocene extinctions, any such doubts have been effectively dispelled for the first human colonizations of oceanic islands in more recent times. For example, about a dozen species of the giant flightless birds known as moas (**Fig. 11.1**), which were relatives of the ostrich and emu, have been described from subfossil bones in New Zealand. There was a long debate about whether the moas died out before or after the Maori colonization in about AD 1000, and hence whether their extinction was due to human or to natural causes. The argument is now settled. Bones of almost all moa species are found in close association with human activity, such as Maori ovens and butchering sites, and the estimated number of moa skeletons exceeds 100,000. Radiocarbon dates indicate that the moas became extinct within about 500 years of the first arrival of humans. There can thus be no reasonable doubt that the moa mass extinction was the result of human activity. Many other New Zealand bird species as well as lizards and frogs became extinct at the same time, probably because of either overhunting or habitat destruction, or the introduction of predatory mammals that arrived with the Maoris.

Fig. 11.1
The moa of New Zealand, a
giant flightless bird related to
the emu and ostrich, which was
rendered extinct by Maori
immigrants.

Like New Zealand, Madagascar formerly had giant flightless birds (known as elephant birds), as is indicated by many
subfossil bones and especially by eggshells. Archaeologists
have found the bones of about a dozen large species of lemur
up to the size of a gorilla, together with those of giant land
tortoises and a hippopotamus. Madagascar suffered the
extinction of this megafauna soon after humans first arrived

around AD 500. One should not overlook the fate of the dodo in neighbouring Mauritius in more recent times. Hawaii has revealed more than 50 species of now-extinct subfossil birds at sites postdating the first colonization by Polynesians; they include especially interesting flightless geese. This mass extinction removed over half of Hawaii's original avifauna. There are similar archaeological reports indicating the former presence of now-extinct birds in many other Pacific islands, including Tonga, Tahiti, Fiji, and New Caledonia. All these islands were not initially occupied by humans, and the indigenous bird populations collapsed soon after their arrival, because they were not adapted to survive the activities either of human hunters or of their introduced predators such as cats and rats. It has indeed been estimated that about one-fifth of all the bird species that existed in the entire world a few thousand years ago have disappeared as a result of human activities on oceanic islands. These recent extinctions of 'naïve' island species, that is, species lacking experience of humans, are now widely accepted as caused by humans, for three reasons: the extinctions are recent enough to be accurately dateable; the close coincidence in time with the arrival of humans; and the lack of evidence for any natural environmental change even remotely adequate to account for the extinctions.

Humans have lived on some tropical Pacific islands (the Bismarcks, the Solomons) for the past 30,000 years, and on the rest of Oceania for periods ranging from 1000 to 3000 years. Their activities have led to the loss of as many as 2000 species of birds that probably otherwise would exist today. This extinction event is documented by fossil birds found at archaeological and palaeontological sites on nearly 70 islands in 19 island groups. The extinction of birds in Oceania rivals the late Pleistocene loss of large mammals in North America as the best-substantiated rapid-extinction episode in the vertebrate fossil record. Some avian extinctions in Oceania occurred within a century or less after human arrival, whereas others required millennia or even tens of millennia. Any of these time-frames is rapid in an evolutionary or a geochronological sense.

Extinctions can be caused by three distinct types of human activity. The first and most obvious is overhunting. For example, about fifty species of moas and large mammals have been exterminated since 1600; and it is very probable that many others were similarly exterminated in the late Pleistocene. Improved hunting technology is rapidly reducing populations of surviving large mammal species, except where they are specially protected. The obvious candidates for the future are most of the large mammals of Africa and south-east Asia living outside game parks and zoos. The second extinction mechanism is the introduction of other mammal species, whether as predators, grazers, browsers, competitors, or vectors of disease. Cats and rats have been the dominant cause of bird extinctions on oceanic islands. Goats have been the main cause of the reduction of island vegetation. The indigenous island birds have evolved no behavioural defences against island predators, and island plants have evolved no chemical or mechanical deterrents to protect them from mammalian herbivores. The third type of extinction mechanism is the destruction of habitats.

The decimation of Australian small marsupials is recent enough for some native species that are still surviving to be greatly at risk in the future, especially from feral cats. We may value cats in the West as household pets, but in the wild they are superbly efficient predators. To many an Australian living in the bush, the only good cat is a dead cat. The largest wave of extinction or risk of extinction in very modern times is probably that of species of indigenous fish, the cichlids, in Lake Victoria, as a result of the predatory Nile perch, which was introduced in a sadly misguided attempt to establish a new fishery.

Whereas the first type of human-induced mass extinction has been checked, if not eliminated, today, by the introduction of laws and regulations to protect the vulnerable organisms, the third type, the destruction of habitats, continues apace and at an accelerating rate. This is most obvious for the most species-rich habitats on land, the tropical rainforests. By the middle of this century, extensive stretches of this habitat are unlikely to survive outside Amazonia and the Congo basin. Each extinction is likely to cause a cascade of secondary extinctions. For example, the human removal of top predators from Barro Colorado Island in Panama caused a population surge in medium-sized predators (monkeys, coatamundis) on which the top predators normally prey. This surge led to the extinction of ground-nesting birds, on which humans had no direct effect. The Chinese are as anxious as anyone to save from extinction their 'flagship' species the Giant Panda, but the small surviving wild population continues its apparently inexorable decline as the extent of its natural habitat, the bamboo forests of western Sichuan, becomes progressively more limited as agricultural demands increase.

Extinctions in the sea

Life on land may be as much as 25 times more diverse than in the sea, a major reason being the diversity of insects. Marine life is thus likely to be correspondingly more vulnerable to extinction. This is in sharp contrast to what the great T. H. Huxley, an extremely well-informed biologist, thought as recently as 1883, when he pronounced that the fishing resources of the oceans were inexhaustible. Even by early in the twentieth century this pronouncement was appearing questionable.

On the basis of an extensive review of palaeoecological, archaeological, and historical data Jeremy Jackson and his many American colleagues have put forward the hypothesis that humans have been disturbing marine ecosystems since they first learned to fish. Extinction by overfishing outstrips all other pervasive human disturbance, including pollution, degradation of water quality, and anthropogenic climate change. Time lags of decades to centuries occur between the onset of overfishing and consequent changes in ecological communities, because unfished species of similar trophic (feeding) level assume the ecological role of overfished species until they too are overfished or die of epidemic diseases owing to overcrowding. Turtles, whales, dugongs, manatees, cod, swordfish, sharks, and rays were formerly very abundant in most coastal ecosystems. The same is true of many oysters and clams, which were once so abundant as to pose hazards to navigation. Their record is now composed of middens, often of massive size, dating back to prehistoric times. In more recent history, oysters were so abundant in British coastal regions in the nineteenth century as to constitute food for the poor; now they count as a luxury, despite their being widely farmed. Severe overfishing drives species to ecological extinction because overfished populations no longer interact significantly with other species in the community. Most ecological research is based on only a few years' study, and so fails to encompass even important environmental disturbances such as the El Niño Southern Oscillation (ENSO), the effects of which are properly appreciated only in the context of decades. This general conclusion derives from detailed studies in a number of different coastal marine habitats.

1. Kelp forests

Kelp forests are shallow, rocky habitats within the photic zone ranging from warm-temperate to subpolar regions worldwide, characterized by kelp (brown algae). In the northern hemisphere kelp forests have experienced widespread reductions in energy supply from other levels (trophic levels) in the food chain and from deforestation resulting from population explosions of herbivores after the removal of apex (top) predators by fishing. Thus removal of the predators of sea urchins (echinoids) results in intense grazing by these common benthic creatures. Consider, for example, the kelp-forest ecosystems distributed along the Pacific margin of northern North America. These include sea otters with the kelp and sea urchins, together with the now-extinct Steller's sea cows, which were widely distributed across the North Pacific rim through the late Pleistocene. These large mammals may have been eliminated from most of their range by aboriginal hunting at the end of the Pleistocene and early Holocene, because they survived thousands of years longer in the western Aleutians that were not peopled until about 4000 years BP. By the time of European contact in 1741, sea cows persisted solely in the Commander Islands, which were then the only islands of the Aleutions unoccupied by aboriginals. European fur traders killed the last sea cow in 1768.

The northern Pacific kelp forests flourished before human settlement because predation by sea otters on sea urchins prevented the sea urchins from overgrazing the kelp. The aboriginal Aleuts greatly diminished the sea-otter populations from about 2500 years BP, with a concomitant increase in size of the sea-urchin populations. Fur traders subsequently hunted otters to the brink of extinction in the 1800s, with the attendant collapse of the kelp forests, which were grazed away by sea urchins released from other predation. Today there is legal protection for the otters, but they suffer increased predation by killer whales, which have changed their diet because of a drastic decline in the populations of seals and sea lions.

A similar sequence of events has been recognized in the Gulf of Maine. Cod and other large ground fish were voracious predators of sea urchins. From the 1920s onwards, new mechanical fishing technology set off a rapid decline in the numbers and body size of coastal cod, which have in more recent times extended offshore to the Georges Bank, formerly a rich fishing ground. Lobsters, crabs, and sea urchins rose in abundance accordingly. Kelp forests disappeared with the rise of sea-urchin populations, and reappeared only when the urchins were in turn reduced by fish.

2. Coral reefs
Coral reefs are the most structurally complex and taxonomically diverse marine ecosystem that provides the habitat for

tens of thousands of associated fish and invertebrates. As has often been noted, they are a marine analogue of tropical rain forests, and are equally vulnerable to disturbance. Recently, coral reefs have experienced dramatic shifts in species dominance owing to intensified disturbance that began a few centuries ago. The effects are most pronounced in the Caribbean, but also can be recognized along the Great Barrier Reef off Queensland despite the intensive protection established there since the 1970s.

Large species of branching *Acropora*, or staghorn corals, dominated reefs in the tropical Atlantic for at least half a million years, indicating the existence of a stable coral community during a long period of time before humans appeared in the region. However, *Acropora* began to decline dramatically in 1980. This catastrophic mortality was due to the overgrowth of algae that exploded in abundance after the mass mortality resulting from disease of the superabundant sea urchin *Diadema antillarium*, which was the last surviving grazer of these algae.

The Great Barrier Reef has suffered recurrent mass mortality since the 1960s because of spectacular outbreaks of the crown of thorns starfish *Acanthaster planci*, which feeds on corals. The cause of these outbreaks remains controversial, but it is almost certainly a new phenomenon. There are no early records of *Acanthaster* in undisturbed fossil deposits or in aboriginal folklore. The outbreaks are now more chronic than episodic. One possible explanation is the overfishing of species that prey on the larval or juvenile stages of *Acanthaster*. Commercial and recreational fishing, as well as the indirect effect of intensive trawling for prawns, are the most likely explanation for the decreased abundance of *Acanthaster* predators. Since the eighteenth and nineteenth centuries the decline in sea cucumbers, whales, dugongs, turtles, and pearl oysters is directly attributable to human exploitation; all five groups have failed to regain their former abundance.

3. Tropical and subtropical sea-grass 'meadows'

The tropical and subtropical sea habitat covers vast areas of tropical and subtropical bays, lagoons, and continental shelves. It formerly contained enormous numbers of sirenian mammals (dugongs, manatees) and turtles, together with a rich diversity of fish and invertebrates. Like coral reefs, they had seemed to be highly resilient to human disturbance until recent decades, when mass mortality of sea grasses became common and widespread. The proximate causes, related ultimately to human activity, appear to be increases in sedimentation, turbidity, and disease.

Vast populations of the very large green turtles were eliminated from off the Americas before the nineteenth century, and from Moreton Bay, Queensland, in the early twentieth century. All turtle species continue to decline at unsustainable rates along the Great Barrier Reef today. Abundant green turtles closely crop turtle grass and greatly reduce the flow of organic matter and nutrients to the underlying sediments. Deposition within the sediment of greatly increased amounts of plant detritus fuels microbial populations, thereby increasing the demand for oxygen and a consequent lack in the oxygen supply (hypoxia or dysoxia) to the water above. All these factors are linked to the recent dying-off of turtle grass from the very well-studied Florida Bay.

Manatees and dugongs were extensively fished by the aborigines and early European colonists of Australia. Vast numbers of these creatures were reported as recently as the end of the nineteenth century, but because of the huge demand for their flesh and oil, the dugong fishery had crashed by the early twentieth century. Dugongs graze extensively. The decline of sea grass in Moreton Bay was almost certainly due to decline of water quality due to eutrophication and runoff of sediments. However the cessation of ploughing of the bay floor by dugongs may also have been a major factor.

4. Estuaries

Temperate estuaries worldwide are undergoing profound changes in oceanography and ecology because of human exploitation and pollution, rendering them the most degraded of all marine ecosystems. The effects include increased sedimentation and turbidity, enhanced episodes of hypoxia and anoxia, loss of sea grasses and dominant suspension-feeding bivalves from the benthic populations, together with a general loss of the oyster-reef habitat, eutrophication (the presence of an excess of plant nutrients), and enhanced microbial production. There is a higher frequency of nuisance algal and toxic dinoflagellate blooms, leading to increased fish mortality. Most explanations point to the increased influx from the land of nutrient materials containing nitrogen and phosphorus (for example, from sewers) as causes of the phytoplankton blooms and eutrophication, but the long-term records demonstrate that losses in benthic suspension-feeders predate eutrophication.

The oldest and longest records are derived from sediment cores in Chesapeake Bay and Pamlico Sound on the eastern margin of the United States, and from the Baltic Sea. These records extend back to about 2500 years BP. In Chesapeake Bay, increased sedimentation and burial of organic carbon began in the middle of the eighteenth century, coincident with widespread land clearance for agriculture. The ecological response was that sea grasses and benthic diatoms declined, while planktonic diatoms and other phytoplankton increased. The flux of organic matter to the bay floor increased dramatically, with the concomitant loss of benthic fauna. Anoxia and hypoxia were not widespread until the 1930s. Similar changes took place in the 1950s in the Baltic, with a widespread expansion of the extent of anoxic laminated sediments, and from the 1950s to 1970s in Pamlico Sound.

Vast oyster reefs were once prominent structures in Chesapeake Bay. Despite intensive harvesting by aboriginal and early colonial populations spanning several millennia, it was not until the introduction of mechanical harvesting with dredges in the 1870s that deep channel reefs were seriously affected. After the oyster fishery had collapsed, hypoxia, anoxia, and other symptoms of eutrophication began to occur in the 1930s. Fishing explains the bulk of the decline; disease and decline of water quality were secondary factors. Field experiments in Pamlico Sound have shown that oysters grow well, survive to maturity, and resist disease when elevated above the zone of summer hypoxia, even in the presence of modern levels of eutrophication and pollution. Dense populations of oysters and other suspension-feeding bivalves graze plankton so efficiently that they limit blooms of phytoplankton and eutrophication. Former inhabitants of Chesapeake Bay that were once abundant but are now virtually eliminated include dolphins, manatees, river otters, sea turtles, alligators, giant sturgeon, sharks, and rays.

5. Offshore benthic communities on the continental shelf

A wide range of fish and shellfish, including cod, haddock, halibut, turbot, flounder, plaice, rays, scallops, clams, and oysters have been fished intensively for centuries off Europe and North America, and more recently throughout the world. There are numerous reports of severely depleted fish stocks, and fishing grounds have progressively moved further offshore. Unfortunately, scientific investigation has consistently lagged behind the economic realities of depleted stocks.

To summarize, the overfishing of large vertebrates and shellfish is the first major disturbance to all the coastal ecosystems that has been studied. There are indeed strikingly similar patterns across ecosystems, despite differences in detail. Everywhere the magnitude of losses has been enormous, both in terms of biomass and abundance of large animals, which are now effectively absent from most coastal ecosystems worldwide. These changes predated ecological observations and cannot be understood except by historical analysis. Their timing in the Americas and the northern Pacific margins closely tracks European colonization and exploitation in most instances. However, aboriginal overfishing also had its effects: for example, the decline of sea otters and possibly sea cows in the north-east Pacific thousands of years ago. Overfishing may often be a necessary precondition for eutrophication and outbreaks of disease, as in Chesapeake Bay. The spread of disease is aided by crowding, and oysters are made less fit by stresses like hypoxia. Changes in climate are unlikely to be a primary reason for microbial outbreaks or disease.

Factors such as nutrient loading and eutrophication, hypoxia, disease, and climate change may be synergistic, so that the whole response of the ecosystem may be much greater than the sum of the individual disturbances. It is quite evident that human impacts have been accelerating in their magnitude and in their effects on the rates of change and diversity of processes responsible for change over time.

What of the future?

The sombre picture outlined above should dispel once and for all the romantic idea of the superior ecological wisdom of non-Western and pre-colonial societies. The notion of the noble savage living in harmony with Nature should be despatched to the realm of mythology where it belongs. Human beings have never lived in harmony with nature. If they caused less disturbance in earlier times it is because of the smaller size of populations and the more primitive technology available. As our world population both increases and demands higher living standards (calling for a wider choice of foods and more living space), the pattern of accelerating disturbance to the natural world is likely to continue into the future. The idea of the ocean as the last frontier to be conquered is a highly disturbing one so far as the depletion of fish stocks is concerned. Serious inroads are already being made into the deep-sea fauna as trawling technology continues to 'improve'. Soon whole oceans, not just coastal regions, will be under serious threat. Policing offshore fishing is already proving far more difficult than policing what is happening on land.

A number of oceanic as well as continental shelf ecosystems have now been thoroughly studied. Industrialised fisheries typically reduce community biomass by 80% within 15 years of exploitation. As regards the biomass of large fish predators such as marlin, swordfish, and tuna, it is estimated today as at only 10% of pre-industrial levels. The declines of large predators in coastal regions have now extended through the global ocean.

Only a Panglossian optimist would deny that human activities have brought the Earth to the brink of biotic crisis. It is likely that there will be a major extinction within the foreseeable future, estimated by some to remove between one-third and two-thirds of all living species. Large land mammals are under especial threat because of the large amounts of space they need. The organisms that will adapt best are likely to be the opportunists, small in size, wide in their range of environmental tolerance, and rapid in reproduction (think rats, flies, and weeds!). Genetic diversity within populations will be increasingly under threat as natural habitats become partitioned and progressively more reduced in size. Current rates of extinction will increase as we continue our assault on the most species-rich areas, such as tropical rain forests and coral reefs. According to the most conservative estimate of the eminent American biologist Edward O. Wilson, who has contributed more than anyone to writing about the diverse threats to modern ecosystems, in the 1990s species were becoming extinct at a rate of three per hour, or 27,000 per year. Within 30 years this figure may rise to several hundred species per day, with consequences that are truly terrible to contemplate. For example, every plant species that becomes extinct may take with it as many as 30 species of insects and other animals that depend on it for food.

Can biodiversity and humans indeed prosper in a world in which most diversity will be confined to relatively small reserves? The picture is not entirely gloomy for those who care, although it may already be too late to save many species, even the 'flagship' species such as the Giant Panda. The growth of environmental consciousness in the West is relatively modern, but is so far very encouraging and is certainly to be welcomed, because the increasing sophistication of technology invented by Europeans and Americans within the past few centuries can be regarded as part of the problem. Until quite recently all the products of nature, whether animal, vegetable, or mineral, were treated within the Christian tradition as part of God's bounty, with no inhibitions on exploitation. Today, a high proportion at least of educated people appreciates the value of conservation measures.

A great deal of persuasion and proselytization still needs to be done, however, in the populous and increasingly wealthy Orient. Although highly sophisticated in other ways, the Chinese and Japanese for the most part exhibit far less environmental consciousness than people in the West. It is not fair to say, as some westerners still do, that the Chinese people as a whole will eat anything that moves. This was only ever true of the Cantonese, of whom other Chinese say that they eat anything with four legs except a table! A visit to a large food market in Guangzhou (Canton) is likely to confirm this, because there is a veritable menagerie of caged animals and birds awaiting the pot. Superstitions can be indulged if they are harmless but not if they threaten endangered species. Those ageing Chinese men worried about their flagging sexual performance who are wealthy enough to buy powdered rhinoceros horn or tiger bone can certainly afford Viagra.

As regards the Japanese, the most obvious concern perhaps is that they continue to hunt minke whales, for purposes of 'research', in defiance of world opinion, but I prefer here to dwell on a subtler matter. In contrast to the rest of the world the Chinese and Japanese eat their food with chopsticks. While, however, the Chinese use plastic ones which can, like cutlery, be washed and re-used, the Japanese use wooden chopsticks which are discarded after each meal. This may seem a trivial example, but think of how many Japanese there are and how many meals they eat in a year, and one can see that it amounts to a lot of trees. A very slight change in eating habits would make a not negligible gesture to conservation.

The kinds of measures that can be adopted, and extended, in response to the growing biological crisis have been talked about and written about extensively by conservation authorities, and need be no more than flagged here. There is a huge need for more biodiversity research in the most species-rich environments, notably tropical rain forest and coral-reef habitats. Most tropical countries suffer from both poverty and weak governance, and need both financial assistance and sympathy in enlisting local interest and support. Ill-informed moralizing will have less effect than a hard-nosed pragmatic approach involving sensitive wildlife management in the context of competing economic forces. The current free-for-all in deep-sea fishing must cease.

What would it cost to save the planet's current biodiversity, with emphasis on the protection of large expanses of key habitats? E. O. Wilson has attempted to estimate how much it might cost, with a few 'best judgement' budgets. The price for buying out loggers (a prime source of damage) in the tropical forests of Amazonia, Congo, and New Guinea would total $5 billion. To protect one-tenth of Amazonia from all threats would cost a mere $250 million, an amount equal to the bill for the failed pathfinder probe to Mars, which was sent with the primary purpose of finding new forms of life.

As an illustration of 'silver bullet' conservation strategies, which contrast with across-the-board efforts, Wilson cites biodiversity hotspots, areas with exceptional concentrations of endemic species that are facing exceptional threats of habitat destruction. These 25 localities, covering only 1.4 per cent of the Earth's land surface, contain the last remaining habitats of 44 per cent of the planet's plant species and 35 per cent of its terrestrial vertebrate species. They represent ecosystems that have already lost at least 70 per cent (in many instances 90 per cent) of their original vegetation. Although the hotspot strategy has already attracted $700 million, support remains far short of the $25 billion estimated to complete the efforts. Protection of these hotspots would contribute vastly to relieving the current extinction problem. However, hotspots are only part, and not even half, of the overall challenge. Among the many steps required to safeguard the biosphere, we must push back the deserts, replant the forests, preserve water supplies, reduce pollution, and restore topsoil.

As a geologist I am only too aware that significant climatic change is recognizable throughout pre-human history, but we have to respect the consensus view of informed scientists who accept that at least an important component of the recent phase of global warming is anthropogenically induced by the burning of fossil fuels. There are worrying signs that corals are already suffering badly, especially in the Indian Ocean, because they are as sensitive to marked temperature rise as to fall. There is an ominous warning from the geological past. Global warming seems to be strongly implicated in the biggest mass-extinction event of all, some 250 million years ago.

Notes and suggestions
for further reading

The book by Hallam and Wignall (1997) provides a general source for further literature on mass extinctions. The references given below accordingly concentrate upon more recent literature.

Chapter 1: In search of possible causes of mass extinctions
The phrase cited in the opening paragraph is used as a subtitle for the stimulating book by Raup (1991). A valuable source for information on viral diseases is Crawford (2000).

Chapter 2: Historical background
New translations and interpretations of the primary texts by Cuvier on fossil bones and geological catastrophes are provided by Rudwick (1997). Lyell's uniformitarian doctrine and the reaction to it are considered in Hallam (1989) and the articles in Blundell and Scott (1998). The subject of neocatastrophism is dealt with in the two books by Ager (1973, 1993) and in the books by Huggett (1989) and Palmer (1998). The enormous influence of punctuated equilibrium theory in general culture has been reviewed in Gould's magisterial book *The Structure of Evolutionary Theory*, published in 2002 shortly before his death. The fields that the theory has touched include subjects as diverse as economics, political theory, sociology, history, and literary criticism.

Chapter 3: Evidence for catastrophic organic changes in the geological record

There is a multitude of textbooks on the principles of stratigraphy and facies interpretation, which need not be cited here, but attention needs to be drawn to a well-received biography of Arthur Holmes by Lewis (2000). For a popular account of cladistics, see Tudge (2000).

Chapter 4: Impact by comets and asteroids

The classic paper that stimulated massive interest in bolide impact as a cause for mass extinctions is by Luis Alvarez *et al.* (1980). Walter Alvarez (1997) followed this up later with a popular book on the impact theory leading up to the discovery of an impact crater in the Yucatán Peninsula of Mexico. Theoretical calculations by Pope (2002), coupled with observations of the coarse dust fraction, indicate that very little of the globally distributed Chicxulub ejecta layer was of submicrometre-size dust. The global mass and grain-size distribution of the clastic debris indicate that stratospheric winds spread the debris from North America across the Pacific Ocean to Europe, and little debris reached high southern latitudes. These findings indicate that the original Cretaceous–Tertiary impact extinction hypothesis – the shutdown of photosynthesis by dust of submicrometre size – is not valid, because it requires more than two orders of magnitude more fine dust than is estimated by Pope. The significant decline in dinosaur diversity within the last two stages of the Cretaceous in the North American Western Interior is recorded in MacLeod *et al.* (1997). The latest Chinese work suggesting survival of the dinosaurs into the early Palaeocene is by Zhao *et al.* (2002).

The recent fullerene research on the Permian–Triassic boundary is published by Becker *et al.* (2001), and the work by the Japanese geochemists on the same boundary is by Kaiho *et al.* (2001). The Becker *et al.* work is contested in *Science* **293**, 2345 (2001); the full text is available on the internet. The Kaiho *et al.* work is challenged by Koeberl *et al.* (2002). The statistical analysis on the Meishan Permian–Triassic section is by Jin *et al.* (2000). Mossman *et al.* (1998) conducted the mineralogical study on quartz at the Triassic–Jurassic boundary in the Newark Supergroup. The abundant planar features found in the quartz grains could readily be related to tectonic activity but not to impact. Iridium anomalies in the late Eocene have been reported by Montanari *et al.* (1993). The Morokweng impact structure in South Africa is described by Koeberl *et al.* (1997).

Chapter 5: Sea-level changes

A general account of sea-level changes in the Phanerozoic is provided by Hallam (1992). Gurnis *et al.* (1998) write about the vertical motions of Australia with respect to the rest of the world in Cretaceous times. The classic paper first exploring the possible relationship between mass extinctions and marine regressions is by Newell (1967). Hallam and Wignall (1999) give a modern review of the relationship between sea-level changes and mass extinctions. The modern work on the Permian–Triassic boundary in East Greenland, relating continental and marine extinctions and giving an estimate of the duration of the extinctions, is by Twitchett *et al.* (2001).

Chapter 6: Oxygen deficiency in the oceans

Useful references for this chapter are the articles in Tyson and Pearson (1991) and the book by Wignall (1994). Wignall and Twitchett (2002) provide the fullest account of extent, duration, and nature of what they describe as the Permian–Triassic super-anoxic event. Numerical experiments by Hotinski *et al.* (2001) indicate that a low equator-to-pole temperature gradient could have produced weak oceanic circulation and widespread anoxia in the ocean at this time. To a first approximation, stagnation would have severely reduced the upwelling supply of nutrients to the photic zone, thereby reducing productivity of surface waters.

A. B. Smith *et al.* (2001) challenge the reality of the Cenomanian–Turonian boundary extinction event in England. This paper has an important bearing on the reality of some other lesser extinction events.

Chapter 7: Climate change

Stanley's (1986) book promotes a climatic cooling model to account for mass extinctions. The ice-age controversy stimulated by Agassiz in the early nineteenth century is reviewed in Hallam (1989). Coope's research on Quaternary beetles is reviewed in Coope (1979). The most recent work on the possible relationship of late Cenozoic extinctions and nutrient deficiency in the western Atlantic is by Allmon (2001). With regard to late Cenozoic climate changes in Africa involving aridification, and an account and discussion of Vrba's turnover pulse hypothesis and the question of human origins, see Vrba *et al.* (1995) and Andrews (1992). The likely relationship between tectonic uplift and climate change in the Cenozoic is discussed in the book by Ruddiman (1997). For the story of methane release from methane hydrate at the Palaeocene–Eocene boundary, see Katz *et al.* (1999).

Retallack (1999) gives an account of climate changes across the Permian–Triassic boundary in Australia, inferred from fossil soils; Smith and Ward (2001) write about the relationship between mass extinction of land vertebrates in South Africa and increased aridification. On a basis of a wide-ranging survey Rees (2002) argues against catastrophic extinction of plants at the end of the Permian, but his results show a very drastic change in the Gondwana Province, where *Glossopteris* was wiped out. There is also evidence from Twitchett *et al.* (2001) of catastrophic elimination of flora in Greenland. Changes in stomatal density among plants across the Triassic–Jurassic boundary are reported by McElwain *et al.* (1999). Hesselbo *et al.* (2000) report on the presumed release of methane from methane hydrate in the Toarcian. A good general account of climate change through the Phanerozoic is provided by Frakes *et al.* (1992).

Chapter 8: Volcanic activity

The information about catastrophic volcanic events in modern and recent historical times is derived from various sources, but principally from the popular book by Officer and Page (1993). The relationship between volcanism and mass extinctions is explored by Courtillot (1999) and Wignall (2001). Recent work on extension of the Siberian flood basalt province into the West Siberian Basin is reported by Reichow *et al.* (2002).

Chapter 9: Pulling the strands together

The relationship of the 'big five' to other mass extinctions in a smooth statistical distribution is discussed by Raup (1994). A good popular account of the largest extinction event of all, at the end of the Permian, has recently been published by Benton (2003). Patterns of generic extinction in the fossil record are considered by Raup and Boyajian (1988) and the geography of end-Cretaceous marine bivalve extinctions by Raup and Jablonski (1993). For relationships between evolutionary size increase and longevity in Jurassic ammonites and bivalves, see Hallam (1975). Erwin (1998) presents a review about recoveries from mass extinctions; other information presented here is gleaned from a variety of sources. Reviews concerning purported extinction periodicity can be found in Hallam (1989) and Hallam and Wignall (1997); see also Benton 1995. For a further discussion of causes, see Hallam and Wignall (1997, 1999) and Wignall (2001).

Chapter 10: The evolutionary significance of mass extinctions

The Red Queen hypothesis was first put forward by Van Valen (1973) and the Stationary model by Stenseth and Maynard Smith (1984). The terms *displacive* and *pre-emptive competition* were first introduced by Hallam (1990). In this article, and that by Benton (1987) the pre-emptive model is favoured over the displacive model as being more consistent with what is found in the fossil record. The late Palaeozoic turnover of plants is described and discussed by DiMichele and Aronson (1992). Conway Morris's promotion of the ubiquity of evolutionary convergence is presented in his book *Life's Solution* (2003). In opposition to the views of Gould (1989), Conway Morris stresses the contingencies in evolutionary history that are produced by, among other things, mass extinctions. Vermeij (1987) gives his views on the late Mesozoic co-evolution of newly evolved predators and shellfish prey.

The second volume of *The Fossil Record* was compiled by Benton (1993) with the collaboration of numerous specialists. On the basis of these data Benton (1995) goes on to review diversity and extinction through the Phanerozoic. For the recognition of self-similarity of extinction statistics, see Solé *et al.* (1997, 1998); for evolutionary patterns from mass originations and mass extinctions, see Hewzulla *et al.* (1999). General patterns of continental and marine diversity are reviewed by Benton (2001). Self-organized criticality in the context of mass extinctions was first proposed by Kauffman (1995).

Chapter 11: The influence of humans

Although nearly two decades old, the chapters in Martin and Klein (1984) remain a valuable source of information on and discussion of Quaternary extinctions on land. Diamond (1989) provides information on island extinctions in more recent historical times. Steadman and Martin (2003) deal with the extinction of Pacific island birds. Owen-Smith published the paper on his 'keystone herbivore' hypothesis in 1987. The computer simulation by Alroy and the recent information on megafaunal extinctions in Australia by Roberts *et al.* were both published in 2001. Jackson *et al.* (2001) review historical overfishing leading to the collapse of coastal ecosystems. Myers and Worm (2003) review the rapid worldwide depletion of predatory fish communities. General information on modern biodiversity and extinction rates is provided by Wilson (1992, 2002), Raven (1999), and Lawton and May (1995).

Bibliography

Ager, D. V. (1973). *The nature of the stratigraphical record.* Macmillan/Wiley, New York.

Ager, D. V. (1993). *The new catastrophism.* Cambridge University Press, Cambridge.

Allmon, W. D. (2001). Nutrients, temperature, disturbance and evolution: a model for the late Cenozoic marine record of the western Atlantic. *Palaeography, Palaeoclimatology, Palaeoecology,* **166**, 9–26.

Alroy, J. (2001). A multispecies overkill simulation of the end-Pleistocene megafaunal mass extinction. *Science,* **292**, 1893–6.

Alvarez, L. W., Alvarez, W., Asaro, F., and Michel, H. V. (1980). Extraterrestrial cause for the Cretaceous–Tertiary extinction: experimental results and theoretical interpretation. *Science,* **208**, 1095–108.

Alvarez, W. (1997). *T. rex and the crater of doom.* Princeton University Press, Princeton, NJ.

Alvarez, W., Asaro, F., and Montanari, A. (1990). Ir profile for 10 million years across the Cretaceous–Tertiary boundary at Gubbio (Italy). *Science,* **250**, 1700–2.

Andrews, P. (1992). Evolution and environment in the Hominoidea. *Nature,* **360**, 641–6.

Becker, L., Poreda, R. J., Hunt, A. G., Bunch, T. E., and Rampino, M. (2001). Impact event at the Permian–Triassic boundary: evidence from extraterrestrial noble gases in fullerenes. *Science,* **291**, 1530–3.

Benton, M. J. (1987). Progress and competition in macroevolution. *Biological Reviews*, **62**, 305–38.

Benton, M. J. (ed.) (1993). *The fossil record*, Vol. 2. Chapman & Hall, London.

Benton, M. J. (1995). Diversification and extinction in the history of life. *Science*, **268**, 52–8.

Benton, M. J. (2001). Biodiversity on land and in the sea. *Geological Journal*, **36**, 211–30.

Benton, M. J. (2003). *When life nearly died: the greatest mass extinction of all time*. Thames and Hudson, London.

Blundell, D. J. and Scott, A. C. (eds) (1998). *Lyell: The past is the key to the present*. The Geological Society, London.

Conway Morris, S. (2003). *Life's solution: inevitable humans in a lonely universe*. Cambridge University Press, Cambridge.

Coope, G. R. (1979). Late Cenozoic fossil Coleoptera: evolution, biogeography, and ecology. *Annual Reviews of Ecology and Systematics*, **10**, 247–67.

Courtillot, V. (1999). *Evolutionary catastrophes: the science of mass extinctions*. Cambridge University Press, Cambridge.

Crawford, D. H. (2000). *The invisible enemy: a natural history of viruses*. Oxford University Press, Oxford.

Diamond, J. M. (1989). The present, past and future of human-caused extinctions. *Philosophical Transactions of the Royal Society of London*, **B325**, 469–77.

DiMichele, W. A. and Aronson, R. B. (1992). The Pennsylvanian–Permian vegetational transition: a terrestrial analogue to the onshore–offshore hypothesis. *Evolution*, **46**, 807–24.

Erwin, D. H. (1998). The end of the beginning: recoveries from mass extinctions. *Trends in Ecology and Evolution*, **13**, 344–9.

Frakes, L. A., Francis, J. E., and Syktus, J. J. (1992). *Climate modes of the Phanerozoic*. Cambridge University Press, Cambridge.

Gould, S. J. (1989). *Wonderful life: the Burgess Shale and the nature of history.* Norton, New York.

Gould, S. J. (2002). *The structure of evolutionary theory.* Harvard University Press, Cambridge, MA.

Gurnis, M., Müller, R. D., and Muresi, L. (1998). Cretaceous vertical motion of Australia and the Australian–Antarctic discordance. *Science*, **279**, 1499–504.

Hallam, A. (1975). Evolutionary size increase and longevity in Jurassic bivalves and ammonites. *Nature*, **258**, 193–6.

Hallam, A. (1989). *Great geological controversies.* (2nd edn). Oxford University Press, Oxford.

Hallam, A. (1990). Biotic and abiotic factors in the evolution of early Mesozoic marine molluscs. In *Causes of evolution – a paleontological perspective* (ed. R. M. Ross and W. D. Allmon), pp. 249–69. University of Chicago Press, Chicago.

Hallam, A. (1992). *Phanerozoic sea-level changes.* Columbia University Press, New York.

Hallam, A. and Wignall, P. B. (1997). *Mass extinctions and their aftermath.* Oxford University Press, Oxford.

Hallam, A. and Wignall, P. B. (1999). Mass extinctions and sea-level changes. *Earth-Science Reviews*, **48**, 217–58.

Hesselbo, S. P., Gröcke, D. R., Jenkyns, H. C., Bjerrum, C. J., Farrimond, P., Morgans Bell, H. S., and Green, O. R. (2000). Massive dissociation of gas hydrate during a Jurassic oceanic anoxic event. *Nature*, **406**, 392–5.

Hewzulla, D., Boulter, M. C., Benton, M. J., and Halley, J. M. (1999). Evolutionary patterns from mass originations and mass extinctions. *Philosophical Transactions of the Royal Society of London*, **B354**, 463–9.

Hotinski, R. M., Bice, K. L., Kump, L. R., Najjar, R. G., and Arthur, M. A. (2001). Ocean stagnation and end-Permian anoxia. *Geology*, **29**, 7–10.

Huggett, R. (1989). *Cataclysms and Earth history.* Clarendon Press, Oxford.

Jackson, J. B. C. *et al.* (2001). Historical overfishing and the recent collapse of coastal ecosystems. *Science*, **293**, 629–37.

Jin, Y. G., Wang, Y., Wang, W., Shang, Q. H., Cao, C. Q., and Erwin, D. H. (2000). Pattern of marine mass extinction near the Permian–Triassic boundary in South China. *Science*, **289**, 432–6.

Kaiho, K. *et al.* (2001). End-Permian catastrophe by a bolide impact: evidence of a gigantic release of sulfur from the mantle. *Geology*, **29**, 815–18.

Katz, M. E., Pak, D. K., Dickens, G. R., and Miller, K. G. (1999). The source and fate of massive carbon input during the latest Paleocene thermal maximum. *Science*, **286**, 1531–3.

Kauffman, S. (1995). *At home in the universe: the search for laws of complexity.* Oxford University Press, New York.

Koeberl, C., Armstrong, R. A., and Reimold, W. U. (1997). Morokweng, South Africa: a large impact structure of Jurassic–Cretaceous boundary age. *Geology*, **25**, 731–4.

Koeberl, C., Gilmour, I., Reimold, W. U., Claeys, P., and Ivanov, B. (2002). End-Permian catastrophe by bolide impact: evidence of a gigantic release of sulfur from the mantle: comment and reply. *Geology*, **30**, 855–6.

Lawton, J. H. and May, R. M. (eds) (1995). *Extinction rates.* Oxford University Press, Oxford.

Lewis, C. (2000). *The dating game.* Cambridge University Press, Cambridge.

MacLeod, N. *et al.* (1997). The Cretaceous–Tertiary biotic transition. *Journal of the Geological Society of London*, **154**, 265–92.

McElwain, J. C., Beerling, D. J., and Woodward, F. I. (1999). Fossil plants and global warming at the Triassic–Jurassic boundary. *Science*, **285**, 1386–90.

Martin, P. S. and Klein, R. G. (eds) (1984). *Quaternary extinctions: a prehistoric revolution.* University of Arizona Press, Tucson.

Montanari, A. (1993), Asaro, F., Michel, H. V. , and Kennett, J. P. (1993). Iridium anomalies of Late Eocene age at Massignano (Italy) and ODP Site 689B (Maud Rise, Antarctic). *Palaios*, **8**, 420–38.

Mossman, D. J., Grantham, R. G., and Langenhorst, F. (1998). A search for shocked quartz at the Triassic–Jurassic boundary in the Fundy and Newark basins of the Newark Supergroup. *Canadian Journal of Earth Sciences*, **35**, 101–9.

Myers, R. A. and Worm, B. (2003). Rapid worldwide depletion of predatory fish communities. *Nature*, **423**, 280–83.

Newell, N. D. (1967). Revolutions in the history of life. *Geological Society of America Special Paper* **89**, 63–91.

Officer, C. B. and Page, J. (1993). *Tales of the Earth: paroxysms and perturbations of the Blue Planet*. Oxford University Press, New York.

Olsen, P. E. *et al.* ((2002). Ascent of dinosaurs linked to an iridium anomaly at the Triassic–Jurassic boundary. *Science*, **296**, 1305–7.

Owen-Smith, N. (1987). Pleistocene extinctions: the pivotal role of megaherbivores. *Paleobiology*, **13**, 351–62.

Palmer, T. (1998). *Controversy, catastrophism and evolution*. Kluwer Academic/Plenum, Dordrecht, New York.

Pope, K. O. (2002). Impact dust not the cause of the Cretaceous–Tertiary mass extinction. *Geology*, **30**, 99–102.

Raup, D. M. (1991). *Extinction: bad luck or bad genes?* Norton, New York.

Raup, D. M. (1994). The role of extinction in evolution. *Proceedings of the National Academy of Sciences*, **91**, 6758–63.

Raup, D. M. and Boyajian, G. E. (1988). Patterns of generic extinction in the fossil record. *Paleobiology*, **14**, 109–25.

Raup, D. M. and Jablonski, D. (1993). Geography of end-Cretaceous marine bivalve extinctions. *Science*, **260**, 971–3.

Raven, P. H. (1999). *Plants in peril: what should we do?* Missouri Botanic Garden, St. Louis, MO.

Rees, P. McA. (2002). Land-plant diversity and the end-Permian extinction. *Geology*, **30**, 827–30.

Reichow, M. K. *et al.* (2002). $^{40}Ar/^{39}Ar$ dates from the West Siberian Basin: Siberian flood basalt province doubled. *Science*, **296**, 1846–9.

Retallack, G. J. (1999). Postapocalyptic greenhouse paleoclimate revealed by earliest Triassic paleosols in the Sydney Basin. *Bulletin of the Geological Society of America*, **111**, 52–70.

Roberts, R. G. *et al.* (2001). New ages for the last Australian megafauna: continent-wide extinction about 46,000 years ago. *Science*, **292**, 1888–92.

Ruddiman, W. F. (ed.) (1997). *Tectonic uplift and climatic change.* Plenum Publishing, New York.

Rudwick, M. J. S. (1997). *Georges Cuvier, fossil bones and geological catastrophes.* University of Chicago Press, Chicago.

Smith, A. B., Gale, A. S., and Monks, N. E. A. (2001). Sea-level change and rock-record bias in the Cretaceous: a problem for extinction and biodiversity studies. *Paleobiology*, **27**, 241–53.

Smith, R. M. H. and Ward, P. D. (2001). Patterns of vertebrate extinction across an event bed at the Permian–Triassic boundary in the Karroo Basin of South Africa. *Geology*, **29**, 1147–51.

Solé, R. V., Manrubia, S. C., Benton, M. J., and Boks, P. (1997). Self-similarity of extinction statistics in the fossil record. *Nature*, **388**, 764–7.

Solé, R. V., Manurbia, S. C., Pérez-Mercador, J., Benton, M. J., and Bak, P. (1998). Long-range correlations and the fractal nature of macroevolution. *Advances in Complex Systems*, **1**, 255–66.

Stanley, S. M. (1986). *Extinction.* Scientific American Books, New York.

Steadman, D. W. and Martin, P. S. (2003). The late Quaternary extinction of birds on Pacific islands. *Earth-Science Reviews*, **61**, 133–47.

Stenseth, N. C. and Maynard Smith, J. (1984). Coevolution in ecosystems: Red Queen or stasis? *Evolution*, **38**, 870–80.

Tudge, C. (2000). *The variety of life*. Oxford University Press, Oxford.

Twitchett, R. J., Looy, C. V., Morente, R., Visscher, H., and Wignall, P. B. (2001). Rapid and synchronous collapse of marine and terrestrial ecosystems during the end-Permian biotic crisis. *Geology*, **29**, 351–4.

Tyson, R. V. and Pearson, T. H. (eds) (1991). *Modern and ancient continental shelf anoxia*. The Geological Society, London.

Van Valen, L. M. (1973). A new evolutionary law. *Evolutionary Theory*, **1**, 1–30.

Vermeij, G. J. (1987). *Evolution and escalation*. Princeton University Press, Princeton, NJ.

Vrba, E. S., Denton, G. H., Partridge, T. C., and Durckle, L. H. (eds) (1995). Paleoclimates and evolution, with emphasis on human origins. Yale University Press, New Haven, CT.

Wignall, P. B. (1994). *Black shales*. Oxford University Press, Oxford.

Wignall, P. B. (2001). Large igneous provinces and mass extinctions. *Earth-Science Reviews*, **53**, 1–33.

Wignall, P. B. and Twitchett, R . J. (2002). Extent, duration, and nature of the Permian–Triassic superanoxic event. *Geological Society of America Special Paper* **356**, 395–413.

Wilson, E. O. (1992). *The diversity of life*. Harvard University Press, Cambridge, MA.

Wilson, E. O. (2002). *The future of life*. Knopf, New York.

Zhao, Z. *et al.* (2002). A possible causal relationship between extinction of dinosaurs and K/T iridium enrichment in the Nanxiong Basin, South China: evidence from dinosaur eggshells. *Palaeogeography, Palaeoclimatology, Palaeoecology*, **178**, 1–17.

Glossary

Words asterisked in the definitions are explained elsewhere in the glossary.

agnathan fishes primitive fishes lacking biting jaws, which achieved a great diversity of forms and sizes in the Ordovician to Devonian periods.

ammonites extinct cephalopods* with coiled shells which flourished as free-swimming carnivores in the Mesozoic era.

angiosperms the flowering plants, which rose to dominance in the Cretaceous period and are by far the most diverse land plants today.

anoxia (adj. anoxic) signifying an absence of oxygen in the environment.

archaeocyathids a group of calcereous sponges which died out at the end of the early Cambrian times, and formed the oldest organic reefs.

basalt one of the most important igneous rocks, produced by crystallization of a molten magma erupted on to the Earth's surface from the interior. It is characterized especially in being relatively low in silica content, and by the minerals plagioclase and pyroxene.

benthic (adj.) referring to the bottom area of a major water body such as a sea or a lake.

benthos a collective term for the organisms occupying the benthic* zone, usually those living on or in the bottom sediment.

belemnites Mesozoic cephalopods with straight, bullet-shaped internal guards made of calcite.

biostratigraphy the correlation of rock strata using their contained fossils.

biota a collective term for the organisms of a particular ecosystem*.

blastoids a group of Palaeozoic echinoderms*.

bolide a collective term for the extraterrestrial bodies asteroids and comets.

brachiopods a phylum of invertebrates characterized by a bivalved calcite shell and distinctive internal structure. Although still extant as minor members of the marine biota they were the dominant benthic* invertebrates of Palaeozoic seas.

bryozoans a group of small colonial invertebrates belonging to a distinct phylum*.

C abbr. for Carboniferous.

catastrophism the doctrine that many events in the geological past were produced by changes of an intensity not observable today.

cephalopods the group of molluscs that includes octopus and squids, as well as the pearly nautilus and the extinct ammonites* and belemnites*.

champsosaurs a Cretaceous–early Tertiary group of aquatic reptiles belonging to the Eosuchia.

chronostratigraphy the correlation of rock strata by their age.

cnidarians the collective name for corals, sea anemones, sea-pens, and jellyfish.

coccolithophorids a group of minute planktonic algae with distinctive calcite skeletons.

conodonts a group of primitive vertebrates with distinctive phosphatic microfossils which flourished in the Palaeozoic and died out at the end of the Triassic period.

corals (rugose and tabulate) two groups of Palaeozoic corals differing significantly in structure from the corals that have existed from the Triassic to the present.

crinoids a group of sessile echinoderms* that survive today in the tropics and subtropics.

Depauperate low in diversity*.

diagenesis (adj.) diagenetic refers to the physical and chemical changes undergone by a sediment in the interval between deposition and consolidation as rock.

diversity (of an ecosystem) is the variety of all the organisms that are present in it.

dysoxia (adj.) dysoxic signifying oxygen deficiency in the environment.

echinoderms members of the phylum* Echinodermata, which includes all the invertebrates with calcite plates and five-rayed anatomy: the starfish, sea urchins, sea lilies, and sea cucumbers.

ecosystem an ecological association of organisms. The term is normally applied only to large-scale phenomena.

eon the largest subdivision of geological time.

epicontinental a term used for marine environments covering the continents as opposed to the oceans. Continents and oceans are underlain by different thicknesses of crust and are fundamentally different from each other. Low-lying continental crust can be covered by shallow sea during times of marine transgression*.

epoch the geological time interval subordinate to a period.

era the geological time interval superior to a period. There are only three; Palaeozoic, Mesozoic, and Cenozoic.

eustatic changes of sea level which are global in extent.

euxinic an aqueous environment, marine or lacustrine, in which the lower part of the water column is anoxic*.

facies the sum total of lithological and palaeontological characteristics of a sedimentary rock. 'Facies' can be used in different senses but usually refers to the inferred former environment.

F–F (abbr. For Frasnian–Famennian).

Flood basalt basalts* erupted as sheet lavas, usually occupying a considerable surface area.

Foraminifera, forams a group of minute single-celled animals with calcite shells that live on the sea bed and in the plankton*, and have a record extending back into the Palaeozoic era.

formation (geological) a key member of the hierarchy of subdivision of rock strata, sometimes defined as the smallest mappable unit.

fusulinids a group of Late Palaeozoic foraminifera* characterized by cylindrical or ovoid shape.

glacioeustatic sea-level changes produced by the melting and freezing of polar ice.

gradualism geological or evolutionary changes produced by gradual processes taking place at a relatively uniform rate.

graptolites a group of colonial planktonic* organims that flourished in the Ordovician and Silurian periods.

gymnosperms 'naked seed' plants, as opposed to the angiosperms* whose seeds are enveloped by soft tissue; includes the conifers and cycads among extant plants.

hadrosaurs a group of Cretaceous dinosaurs.

horizon (geological) a level in a succession of rock strata that can be characterized by a particular fossil or other distinctive feature.

ichthyosaurs a group of Mesozoic reptiles secondarily adapted to an active marine swimming life, with a broad resemblance to dolphins.

inoceramids a group of benthic* marine bivalve molluscs that flourished in the Cretaceous period.

J abbr. for Jurassic.

K abbr. for Cretaceous.

K strategy an evolutionary strategy adopted by organisms which entails high investment in preserving a small number of young, and is characteristically associated with growth to a relatively large size for individual members of the group.

Lazarus taxa organisms that survive mass extinctions in small numbers and return to the fossil record after a significant temporal absence.

lithistids a group of sponges with siliceous spicules.

lithostratigraphy the correlation of strata using the lithological characteristics of the rocks.

macroinvertebrates invertebrates readily visible to the naked eye.

marsupials mammals characterized by possession of a birth pouch and lacking a placenta, e.g. kangaroos, opossums.

micropalaeontology the study of microscopic fossils.

mosasaurs a group of large-sized, marine carnivorous swimming lizards that flourished for a brief time in the late Cretaceous.

nanoplankton planktonic organisms so minute that they can be adequately studied only by using electron microscopy.

nautiloids a group of shelled cephalopods including the extant pearly nautilus. Unlike this animal, the great majority of Palaeozoic nautiloids did not have plane-spiral shells.

nektobenthos organisms swimming close to the sea bed.

nekton organisms swimming within the water column, between the plankton* and the benthos*.

neocatastrophism a term used to designate a late twentieth-century revival of the old doctrine of catastrophism*.

out-compete (verb) refers to organisms relatively successful in competing with other organisms, e.g. for food resources and territory.

oxic (adj.) signifying that oxygen is abundantly present in the environment.

P abbr. for Permian.

palaeontology the study of fossils.

palynology the study of pollen and spores.

pelagic (adj.) of the open ocean as opposed to inshore waters. Sometimes confused with planktonic*.

period (geological) the geological interval of time between era* and epoch*.

photic zone that part of oceanic waters into which light penetrates.

phylum (pl. phyla) the largest taxonomic subdivision of major groups of organisms such as animals and plants.

phytoplankton those members of the plankton* that do not include animals, essentially plants and algae that photosynthesize.

placoderms a group of heavily armoured fish that lived mainly in the Devonian period.

plankton (adj.) planktonic organisms living in the surface waters of oceans or lakes.

plesiosaurs a group of Mesozoic marine swimming carnivorous reptiles.

pseudoextinction apparent extinction caused by the evolution of one group of organisms into another and thus signified by a change of fossil name.

pterosaurs a group of Jurassic and Cretaceous flying reptiles comparable to bats.

radiation (evolutionary) expansion or diversification, perhaps as a result of a group speciating rapidly, as in exploiting an ecological opportunity following a mass extinction, or the acquisition of a new adaptation.

regression (marine) the withdrawal of seas from continental shelves or epicontinental* seas.

r strategy an evolutionary strategy adopted by organisms which involves an unusually high rate of reproduction and early achievement of sexual maturity. It is characteristically associated with small size and physically unstable environments.

rudists a group of large, thick-shelled bivalve molluscs of Jurassic and Cretaceous age.

sauropods a major group of herbivorous dinosaurs.

scablands unusual landforms, as in eastern Washington State, thought to have been produced by catastrophic floods.

sclerosponges sponges with hard calcareous skeletons.

section (geological) the detailed designation of a succession of rock strata in terms of lithological change.

speciation the process of formation of new species from a precursor.

stage (geological) the principal subdivision of rock strata beneath the level of system*.

stegosaurs a group of Jurassic herbivorous dinosaurs.

stratigraphy the scientific study of sedimentary rock successions, comprising both correlation and environmental interpretation.

stratum (pl. strata) the smallest lithologically distinguishable units in sedimentary rock successions.

stromatoporoids calcereous masses of layered and structured material found in limestone successions of Cambrian to Cretaceous age and dominant in the Silurian and Devonian. Palaeozoic stromatoporoids were important reef-formers. They are now thought to be related to sclerosponges*.

symbiotic association an ecological association of organisms involving mutual benefit; e.g. between corals and their dinoflagellate algae.

system the major stratigraphic subdivision. Note that, e.g. the Jurassic system* refers to a succession of rock strata, while the Jurassic period* refers to the time during which they were formed.

T (abbr. for Triassic).

taxon (pl. taxa) the classification of organisms, as in the ascending hierarchy of taxa: species, genus, family, order, phylum.

taxonomy the study of the classification of organisms.

tectonoeustatic (adj.) refers to global sea-level changes produced as a result of tectonic movements of the Earth's crust, e.g. the rise and subsidence of oceanic ridges.

tektites glassy objects formed from terrestrial material melted and displaced by the impact of an extraterrestrial body, such as a meteorite or comet.

teleosts the most evolutionary advanced group of bony fishes, which arose in the late Mesocoic and is dominant today.

thermophilic (adj.) 'heat-loving'.

transgression (marine) the spread of seas over continental shelves or epicontinental* seas.

trilobite a major group of arthropods, with three-lobed bodies and many pairs of limbs, which dominated early Palaeozoic benthic faunas and became extinct in the Permian period.

type section that section* which has been designated as the type example of, for instance, a stage* or system* boundary.

unconformity (adj. unconformable) a break in the stratigraphic record, representing a period of time not marked by sediment deposition. Sometimes there is an angular discordance between the older strata beneath the unconformity surface and the younger strata above. This usually signifies tectonic disturbance produced by stratal folding or faulting.

uniformitarianism the doctrine originally propounded by Charles Lyell, signifying that all past geological events could have taken place by changes no different in rate or style than those that can be observed today.

velociraptors a group of small, active carnivorous dinosaurs of Cretaceous age.

zone a stratigraphic unit characterized by a particular fossil, the zone fossil.

zooplankton animal components of the plankton*.

Index